SAFE USE OF SMART DEVICES IN SYSTEMS IMPORTANT TO SAFETY IN NUCLEAR POWER PLANTS

The following States are Members of the International Atomic Energy Agency:

AFGHANISTAN
ALBANIA
ALGERIA
ANGOLA
ANTIGUA AND BARBUDA
ARGENTINA
ARMENIA
AUSTRALIA
AUSTRIA
AZERBAIJAN
BAHAMAS
BAHRAIN
BANGLADESH
BARBADOS
BELARUS
BELGIUM
BELIZE
BENIN
BOLIVIA, PLURINATIONAL
  STATE OF
BOSNIA AND HERZEGOVINA
BOTSWANA
BRAZIL
BRUNEI DARUSSALAM
BULGARIA
BURKINA FASO
BURUNDI
CAMBODIA
CAMEROON
CANADA
CENTRAL AFRICAN
  REPUBLIC
CHAD
CHILE
CHINA
COLOMBIA
COMOROS
CONGO
COSTA RICA
CÔTE D'IVOIRE
CROATIA
CUBA
CYPRUS
CZECH REPUBLIC
DEMOCRATIC REPUBLIC
  OF THE CONGO
DENMARK
DJIBOUTI
DOMINICA
DOMINICAN REPUBLIC
ECUADOR
EGYPT
EL SALVADOR
ERITREA
ESTONIA
ESWATINI
ETHIOPIA
FIJI
FINLAND
FRANCE
GABON
GEORGIA

GERMANY
GHANA
GREECE
GRENADA
GUATEMALA
GUYANA
HAITI
HOLY SEE
HONDURAS
HUNGARY
ICELAND
INDIA
INDONESIA
IRAN, ISLAMIC REPUBLIC OF
IRAQ
IRELAND
ISRAEL
ITALY
JAMAICA
JAPAN
JORDAN
KAZAKHSTAN
KENYA
KOREA, REPUBLIC OF
KUWAIT
KYRGYZSTAN
LAO PEOPLE'S DEMOCRATIC
  REPUBLIC
LATVIA
LEBANON
LESOTHO
LIBERIA
LIBYA
LIECHTENSTEIN
LITHUANIA
LUXEMBOURG
MADAGASCAR
MALAWI
MALAYSIA
MALI
MALTA
MARSHALL ISLANDS
MAURITANIA
MAURITIUS
MEXICO
MONACO
MONGOLIA
MONTENEGRO
MOROCCO
MOZAMBIQUE
MYANMAR
NAMIBIA
NEPAL
NETHERLANDS
NEW ZEALAND
NICARAGUA
NIGER
NIGERIA
NORTH MACEDONIA
NORWAY
OMAN
PAKISTAN

PALAU
PANAMA
PAPUA NEW GUINEA
PARAGUAY
PERU
PHILIPPINES
POLAND
PORTUGAL
QATAR
REPUBLIC OF MOLDOVA
ROMANIA
RUSSIAN FEDERATION
RWANDA
SAINT KITTS AND NEVIS
SAINT LUCIA
SAINT VINCENT AND
  THE GRENADINES
SAMOA
SAN MARINO
SAUDI ARABIA
SENEGAL
SERBIA
SEYCHELLES
SIERRA LEONE
SINGAPORE
SLOVAKIA
SLOVENIA
SOUTH AFRICA
SPAIN
SRI LANKA
SUDAN
SWEDEN
SWITZERLAND
SYRIAN ARAB REPUBLIC
TAJIKISTAN
THAILAND
TOGO
TONGA
TRINIDAD AND TOBAGO
TUNISIA
TÜRKİYE
TURKMENISTAN
UGANDA
UKRAINE
UNITED ARAB EMIRATES
UNITED KINGDOM OF
  GREAT BRITAIN AND
  NORTHERN IRELAND
UNITED REPUBLIC
  OF TANZANIA
UNITED STATES OF AMERICA
URUGUAY
UZBEKISTAN
VANUATU
VENEZUELA, BOLIVARIAN
  REPUBLIC OF
VIET NAM
YEMEN
ZAMBIA
ZIMBABWE

The Agency's Statute was approved on 23 October 1956 by the Conference on the Statute of the IAEA held at United Nations Headquarters, New York; it entered into force on 29 July 1957. The Headquarters of the Agency are situated in Vienna. Its principal objective is "to accelerate and enlarge the contribution of atomic energy to peace, health and prosperity throughout the world".

SAFETY REPORTS SERIES No. 111

SAFE USE OF SMART DEVICES IN SYSTEMS IMPORTANT TO SAFETY IN NUCLEAR POWER PLANTS

INTERNATIONAL ATOMIC ENERGY AGENCY
VIENNA, 2023

COPYRIGHT NOTICE

© IAEA, 2023

Printed by the IAEA in Austria
January 2023
STI/PUB/1975

IAEA Library Cataloguing in Publication Data

Names: International Atomic Energy Agency.
Title: Safe use of smart devices in systems important to safety in nuclear power plants / International Atomic Energy Agency.
Description: Vienna : International Atomic Energy Agency, 2023. | Series: IAEA Safety Reports Series, ISSN 1020-6450 ; no. 111 | Includes bibliographical references.
Identifiers: IAEAL 22-01514 | ISBN 978–92–0–120122–5 (paperback : alk. paper) | ISBN 978–92–0–120222–2 (pdf) | ISBN 978–92–0–120322–9 (epub)
Subjects: LCSH: Nuclear power plants — Safety measures. | Nuclear power plants —Electronic equipment. | Nuclear power plants — Instruments.
Classification: UDC 621.039.58 | STI/PUB/1975

FOREWORD

Increasingly, the nuclear industry is faced with the need to replace analogue devices that have reached their end of life and become unmaintainable or obsolete, alongside a lack of qualified or qualifiable analogue devices to replace them. Given the small scale of the nuclear market, the nuclear industry is increasingly looking to utilize devices designed for non-nuclear applications, which often contain non-analogue or digital components (so-called smart devices), in systems important to safety in nuclear power plants.

Industrial or commercial grade smart devices are typically developed according to non-nuclear-industry standards. Some of these devices are certified by non-nuclear organizations using those non-nuclear standards in industrial safety applications (e.g. the oil, gas, rail and aircraft industries). The qualification of an industrial or commercial smart device for applications in nuclear power plant systems important to safety may often be more difficult than it would be for a device specifically developed for nuclear applications because the commercial development processes for such devices may be less transparent and controlled than the processes described in the relevant IAEA safety standards. Often, the qualification is challenging if there is no cooperation from the manufacturer. The difficulty associated with the use of these devices may relate to the unavailability of information to demonstrate quality and reliability.

Gaining access to information concerning the design and manufacturing of such devices to enable their evaluation and the implementation of the necessary compensatory measures for their acceptance can be difficult. In addition, the end users or applicants need to have such information to generate evidence for the regulatory body.

Currently, there is limited regulatory consensus on the safe use of smart devices in nuclear safety systems. Safe use entails selecting and evaluating smart devices for use in nuclear power plants, making use of third party certification within the framework of the assessment process and adequately implementing safety design criteria using a graded approach.

The 43rd meeting of the Commission on Safety Standards requested the Secretariat to produce a publication to address current practices for the selection and evaluation of industrial digital devices of limited functionality, including smart devices, to ensure the safe use of such devices in nuclear systems important to safety.

This publication was produced by an international committee of experts and advisors, whose experience and knowledge were valuable in providing a comprehensive technical basis for the development of this report. The IAEA wishes to thank all the participants and their Member States for their contributions. The IAEA officer responsible for this publication was A. Duchac of the Division of Nuclear Installation Safety.

CONTENTS

1. INTRODUCTION

1.1. BACKGROUND

As a result of rapidly advancing digital technologies, smart devices[1] are found in an increasing number of applications in both operating and new nuclear power plants (NPPs). These smart devices can be implemented as separate or stand-alone field components or may be embedded as components in other equipment or systems; they can be used to increase plant reliability, enhance safe operation and improve testing and monitoring functions. However, the use of smart devices may potentially introduce new hazards, vulnerabilities and failure modes.

Smart devices incorporate either software[2] or digital circuit designs created using hardware description language (HDL). They are used in instrumentation and control (I&C) and electrical systems to typically perform limited functions, which are defined by the designer and manufacturer and are usually configurable — but not modifiable or reprogrammable — by the end users. The user configurability for these smart devices is normally limited to parameters relating to compatibility with the process being monitored or controlled, or to interfaces with connected equipment. Smart devices can also be used within electrical power systems (e.g. AC/DC power supplies, adjustable speed drives and digital protection relays) and other systems in NPPs.

Although some smart devices are developed specifically for nuclear safety applications, the introduction of new digital technologies, as well as the small size of the market for specific 'nuclear grade' I&C components, forces designers to use industrial or commercial grade smart devices in various systems important to safety. These smart devices are also used in the maintenance of existing NPP systems, because identical replacements for existing devices are no longer available.

Industrial or commercial grade smart devices are typically developed according to non-nuclear-industry standards (see, for example, International Electrotechnical Commission (IEC) standard 61508 [1]). Some of these smart devices are certified by non-nuclear organizations using non-nuclear standards for use in industrial level (or non-nuclear) safety applications (e.g. the oil, gas, rail and aircraft industries). The qualification of an industrial or commercial

---

[1] There is no simple definition of a smart device; however, smart device characteristics are provided in Section 1.3. Smart devices are generally digital devices of limited functionality and do not include user programmable devices such as programmed logic controllers.

[2] In most cases, the software will have been predeveloped.

grade smart device for application in NPP systems important to safety could be challenging because commercial development processes are often less transparent and controlled than those described in IAEA Safety Standards Series No. SSG-39, Design of Instrumentation and Control Systems for Nuclear Power Plants [2], unless the device has been developed to a recognized safety standard.

Gaining access to information from the manufacturer concerning the design and manufacturing of smart devices to enable their evaluation with a view to accepting them in various systems important to safety can be difficult. In addition, end users or applicants need to have sufficient information to generate evidence that the candidate devices are suitable for applications in systems important to the safety of NPPs.

SSG-39 [2] provides limited, high level guidance on the qualification of digital devices of limited functionality (DDLFs) for use in I&C systems important to safety. In 2013, the IEC published standard IEC 62671 [3] on the selection and use of industrial DDLFs and digital devices of limited configurability for use in NPPs. Additional details on the techniques and measures expected at different safety levels are available in IEC 61508 [1].

1.2. OBJECTIVE

The purpose of this publication is to address safety aspects and criteria associated with the safe use of industrial commercial smart devices in systems important to safety either in individual or multiple applications in an NPP, but it may also be useful in the consideration of devices that have been developed to nuclear standards. These applications may be in I&C, electrical, mechanical and any other systems or stand-alone equipment in an NPP.

Guidance provided here, describing good practices, represents expert opinion but does not constitute recommendations made on the basis of a consensus of Member States.

1.3. SCOPE

This publication considers smart devices that have the following characteristics (adapted from IEC 62671 [3]):

(a) The device is a digital device that contains software or programmed logic (e.g. a hardware programmed device (HPD)) and is a candidate for use in an application important to safety.

(b)   The primary function performed is well defined and applicable to only one type of application within an I&C or electrical system, such as measuring a temperature, pressure or voltage, positioning a valve, controlling the speed of a mechanical device or performing an alarm function.

(c)   The primary function performed is conceptually simple[3] and limited in scope (although the manner of accomplishing this internally may be complex).

(d)   The device is not designed so that it is reprogrammable after manufacturing, nor can the device's functions be altered in a general way so that it will perform a different function; only predefined parameters can be configured by users.

(e)   If the primary device function can be tuned or configured, this capability is restricted to parameters related to the process (such as process range), performance (speed or timing), signal interface adjustment (such as selection of voltage or current range) or gains (such as adjustment of proportional band).

(f)   The device might be stand-alone or embedded as a part of a piece of equipment or system. 'Embedded' refers to the embedding of a low level module within a device or module that may otherwise not contain any 'smart' aspects. The concern is that the embedded device may not be identified, and its failure modes might not be well understood by the supplier.

(g)   The device might not be the only digital device in the plant, thus there is a concern about the potential for common cause failures (CCFs) among smart devices in the plant.

The scope of this Safety Report covers devices that accomplish one primary and fundamentally unchangeable function with minimal or no ancillary functions. If ancillary functions exist in the device, they will be identified and assessed in terms of their potential to interfere with the primary function of the device. Comparing the trends in other markets, these devices will typically have non-interfering functions with only data export features.

---

[3] 'Conceptually simple' is a difficult term to define, since it may refer to anything from performing a simple mathematical transformation, such as extracting a square root (ubiquitous in flow measurement), to limiting the value of a signal within some adjustable band or performing a proportional integral derivative control function. In each case, the scope of the functionality is limited to a single one of such types of function. This definition of simplicity may extend to gas analysers or vibration monitors, as long as these functions are specific to the application and cannot be converted to some other use. On the same basis, stand-alone electrical protection relays, which perform functions such as reverse power trips or undervoltage protection that have traditionally been implemented with analogue technology, may be included under this definition of conceptual simplicity. (Note that networked protection relays are not included.)

The scope of this publication excludes devices that provide the capability to define functionality with either a general purpose language, such as C (or its many derivatives), or an application specific language, such as ladder logic or function blocks. The following devices may serve as examples, assuming that they provide only a limited degree of configurability, as described above:

— Pressure, temperature and other process sensors;
— Smart sensor transmitters (e.g. pressure transmitters);
— Valve positioners;
— Electrical protective devices, such as overvoltage and overcurrent relays;
— Motor starters;
— Variable speed drives;
— Display units (e.g. multisegment light emitting diode bar displays);
— Simple communications interface modules or devices.

It is not possible to list all devices that do not fall within the scope of this Safety Report, but the equipment and devices listed below serve as examples:

— Programmed logic controllers and other platforms of similar capabilities and complexity;
— Devices provided with a programmable language, regardless of its restricted nature (in terms of number of function blocks (or equivalent) or inputs and outputs), that have been designed to be configurable for more than one application (e.g. a single loop digital controller with a function block language).

Smart devices as defined in this report are also limited in functionality. All other devices that use commercial computers (such as personal computers, industrial computers or programmable logic controllers) and are developed for an I&C platform are not included in the scope of this Safety Report.

This publication discusses the following safety aspects related to the development process, selection and evaluation of smart devices:

— Differences in the development process, design intent and qualification process between devices that are developed specifically for the nuclear market and those that are not.
— Motivations for using smart devices in NPPs.
— Potential technical concerns that could be introduced by smart devices. Safety consideration of the safe use of smart devices of limited functionality in systems important to safety, including considerations around technological obsolescence.

— Architectural considerations relating to including smart devices.
— Consideration of CCF in the design of the architecture of systems important to safety. Smart devices include mechanisms that do not exist in analogue devices.
— A process to select and evaluate candidate smart devices for their safe use, which includes I&C, electrical, mechanical and any other systems in NPPs.
— Criteria for evaluating the functional suitability of a smart device that contains predeveloped software or firmware or uses digital circuits designed with HDL.
— Use of third party certification of smart devices developed according to non-nuclear standards and certified by non-nuclear organizations using non-nuclear standards for use in NPP applications.

Finally, this report does not cover the qualification of the mechanical hardware aspects of a smart device and its compatibility with the plant. The term 'hardware' is used to refer to electronic hardware, rather than to the entirety of electromechanical components.

1.4. STRUCTURE

Section 2 defines the scope of the discussion; that is, the motivation and challenges associated with using smart devices in various systems important to safety. Section 3 discusses overall architectural considerations in the use of smart devices in various systems important to safety. Section 4 provides information on the methods used to qualify smart devices for their use in safety applications, the content of the qualification documentation and the methods used for accepting equipment for implementation in its target application in an I&C system. Section 5 provides information on the life cycle of a smart device in its target application in an I&C system of an NPP. Annex I provides some optional additional considerations relative to the selection, qualification and use of smart devices in various systems important to safety. Annex II presents several examples of the analysis of CCFs when using smart devices at different levels in the overall I&C architecture. Annex III provides a list of international and national standards that have a strong relationship with this Safety Report. Annex IV provides examples of a framework and practices for the qualification of smart devices in selected Member States.

2. MOTIVATION AND CHALLENGES ASSOCIATED WITH SMART DEVICES

2.1. ADVANTAGES OF USING SMART DEVICES

The use of smart devices in systems important to safety in NPPs started in the 1980s, motivated by several issues that are becoming more evident with time. Certain aspects particular to smart devices create a number of opportunities and challenges that come with their use. A selection of these is given in Sections 2.1.1–2.1.7.

2.1.1. Solution to obsolescence of analogue devices

The relatively small size of the nuclear market, the gradual obsolescence of the original devices and the advancement of digital technologies have led manufacturers to stop manufacture of older, analogue designs in favour of smart devices, which can offer advantages in large numbers of industrial applications. Devices as conceptually simple as a reverse power relay are now manufactured as smart devices that can communicate using dedicated high security protocols, which can significantly improve the availability of power supply systems.

2.1.2. Implementation of functions requiring many analogue components

Even functions that are conceptually relatively simple may require six or more discrete analogue devices, any of which may contribute to the failure of a safety function, resulting in poor reliability for the overall function. For example, consider a trip set point that varies linearly between two fixed values as a function of the average of three input signals; such a set point function would require at least three, and perhaps more, discrete analogue devices, such as a summer and limiters.

A simple network of such analogue devices would have several times the probability of failure of a single smart device and would not be able to signal a failure. A single smart device can replace several analogue devices, take up less space and provide an alarm or trip upon detection of an internal or input failure, and potentially do this with higher reliability and improved diagnostics.

2.1.3. Additional functionality to maintain safe operation following a failure

In the example presented in Section 2.1.2, the smart device can perform a spread check on the three analogue input devices and substitute either a median selection or a high selection, depending on which will best preserve safety following a failure, and can either open a contact to signal the failure for immediate maintenance or trigger an immediate trip.

2.1.4. Self-diagnostics to detect random failures

Failures of analogue devices may be silent, in the sense that they might not be discovered until the next scheduled surveillance test. Smart devices almost always include self-diagnostics that cover all the dominant failure modes. When these reveal a failure, the device will open a 'fault' contact or output an out-of-range signal, such as defined by the NAMUR NE43 standard [4], which downstream devices will immediately treat as a discovered failure indicating the need to take safety action. This can also trigger immediate maintenance, thereby significantly increasing the availability of the safety function.

2.1.5. Reduced operational and maintenance costs

The ability of smart devices to output recognized 'safe' values inherently detectable by downstream devices, as well as to alert operators and maintenance personnel to failure, potentially saves maintenance costs because of the shortened diagnostic time. The potential for using fallback or substitute signals in the face of a failure also introduces the possibility of continuing to operate under some restrictions, such as reduced power, instead of being forced to shut down immediately. In some locations, this advantage may even be crucially beneficial to public safety by increasing the reliability of the electrical grid.

Despite the potential for reduced operation and maintenance costs, the installation and proliferation of smart devices may mean an evolution of working practices at the plant to which some cost may be attached. Training is often needed for maintenance staff in working with digital devices and the organization of logistics (for example, in the case where a replacement stock of the same device model is used for different applications in different configurations). Furthermore, because of their complexity, such devices may have higher replacement costs than the analogue devices used previously, and maintenance costs might rise as a result of software licensing and frozen version procurement contracts.

2.1.6. Potential reduction of surveillance requirements

The built-in self-diagnostic features of a smart device could be used to potentially reduce some of its surveillance requirements. The extent of the scope for reduction in traditional surveillance testing is based on detailed analysis of the effectiveness of built-in diagnostics in comparison with the equivalency and effectiveness of surveillance testing in the detection of failure modes.

2.1.7. Improved monitoring of mechanical and electrical systems

Smart devices have the potential to apply more improved condition monitoring of large equipment and to provide more refined protection functions. Condition monitoring, such as vibration monitoring and gas detection monitoring, can provide early detection of impending failures. Protection logic using smart devices can use more complex protection limits — such as a trip set point that is the minimum of two limits, such as $f(x)$ and $g(x)$, rather than a constant trip set point — and thereby improve both the operating margin to prevent spurious trips and the margin of safety against equipment damage or plant safety.

2.2. TECHNICAL CHALLENGES REGARDING IMPLEMENTATION OF SMART DEVICES

2.2.1. Internal complexity of the realization of the desired functionality

It can be challenging to demonstrate that complex, programmable digital I&C systems conform with the fundamental design principles of safety I&C systems, such as independence, diversity and defence in depth, redundancy and determinism (predictability and repeatability).

Conventional industrial applications widely use devices containing microprocessors and HPDs (hardware devices with internal circuitry configured using software based tools) that typically feature considerable configurability, including low level programmability. These attributes can enhance the market reach for the industrial market but are generally negative in the nuclear context, where simplicity of both conceptual functionality and internal design are desirable. The aim of this simplicity is to ensure that there is effective understanding of the device's behaviour under all conditions, including failure modes. Consequently, a candidate device for use in systems important to safety needs to be focused narrowly on a single type of use, with at worst limited and preferably no non-essential functionality beyond that purpose.

One difficulty is that even devices specifically designed for performing limited, conceptually simple functions often feature internal complexity. Added complexity associated with the performance of functions not directly related to the main device functions may introduce potential design errors or additional hazards. For example, on-line self-testing and self-diagnostics functions, which are routinely incorporated into programmable digital I&C devices, could improve system availability and reliability, but could also add complexity to the system design.

2.2.2. Extent of configurability

While a device considered to be a smart device cannot be programmable after manufacture, it may be possible for the end user to configure a set of predefined parameters. Normally, the only parameters that can be configured are those related to the process (such as process range), performance (speed or timing), signal interface adjustment (such as selection of voltage or current range) or gains (such as adjustment of the proportional band).

This configurability may enable wide application of the same device or may improve the performance of a safety function by the possibility of non-linear or filtered signal processing, for example. The configurability may also provide a new source of errors that might be introduced, and both thorough review and configuration management (see Section 2.3.7) are required to ensure that configuration changes are understood, safe and preserved during maintenance.

2.2.3. Suitability of internal sampling frequency and frequency response

Analogue devices theoretically have a very wide bandwidth, which means that they are capable of a fast response to plant transients, although this fast response is usually filtered to reduce the susceptibility to noise. Smart devices are subject to built-in sampling times that limit fast response and therefore require different trade-offs in design from analogue devices. In particular, the software for microprocessors that support multiple tasks has to be explicitly designed to guarantee the time response of the primary function for which the smart device will be selected. This is less of a concern with HPDs, since they usually do not contain software (note, however, that some HPDs include on-board microprocessors, and these need to be examined as well during the assessment of the device's suitability).

2.2.4. Potential existence of secondary functionality

Dedicated, specially designed smart devices embedded in other products (for example, to provide protection for motors) are less likely to contain secondary functions, but smart devices that are not embedded in a larger device almost always contain secondary functions. These functions may provide, for example, the human interface needed to set process related parameters such as the pressure range and the filtering time constants. The concerns about secondary functions include the following:

— The potential for interference with the primary functions resulting from a software or hardware fault related to a secondary function;
— The potential for unintended or unapproved changes to operating parameters if it is not possible to lock the device settings;
— The potential that audits might not be able to confirm both the version of software (or its equivalent) and the settings of operational parameters without risk of modifying these items.

2.2.5. Undocumented features

A related issue regarding the use of smart devices in NPPs is the possibility of a device having undocumented or unintended features capabilities. Smart devices may have unintended features because of the increased complexity and flexibility desirable in the non-nuclear market. In the case of a smart device with specific industrial communication protocols, the complexity related to self-diagnostics and the memory management scheme for the protocol stack could result in unintended failure.

Paragraph 2.114 of SSG-39 [2] states:

"Commercial off the shelf devices…may have unintended functionalities… that are not needed in the nuclear power plant application.... The difficulty associated with acceptance of a commercial off the shelf device may often lie with the unavailability of the information to demonstrate quality and reliability."

Lack of sufficient information on how such unintended features affect the devices could lead to a failure mode that is difficult to evaluate or estimate.

Paragraph 2.81 of SSG-39 [2] lists failure mode and effects analysis (FMEA) as a typical design analysis, verification and validation (V&V) technique that is "often used to confirm compliance with the single failure criterion and to confirm that all known failure modes are either self-revealing or detectable by planned testing". Unidentified failures of the smart device may result in undetected failure and increase the uncertainty of system level FMEAs. Furthermore, the potential failure could affect the performance of safety functions or defeat a fail-safe design. Therefore, consideration of the use of smart devices in the design of systems important to safety dictates the need for additional quality and reliability information on smart devices to be specified and maintained.

2.2.6. Necessity of ensuring a secured configuration

It is common for smart devices to be equipped with a human–machine interface unit (either permanently attached (possibly with a built-in display) or plugged in when needed), which provides access to view and modify the configuration parameters of the device. Various levels of access with the appropriate level of control may be required for the following purposes:

— To ensure that the current parameter settings can be confirmed (during audits, for example) without risk of accidentally modifying them (passive access);
— To provide authorized access as needed to perform surveillance testing (low level active access);
— To provide a higher level of secure, privileged access to set parameters crucial to the safety function performed by the smart device (high level active access).

These levels of access may be achieved by a password or a physical key, or a combination of these methods, and would preferably provide a remote indication of the activation of active access.

2.2.7. Radiation susceptibility

Energetic or ionized particles penetrating the sensitive cross-section of a semiconductor may induce a single bit (or multibit) transient in microelectronics components, introducing an unexpected failure mode. Thus, the radiological environment may result in a temporary upset or permanent damage to a smart device with microelectronics.

Digital devices for airborne, military and space applications usually comply with the recommendations of ASTM International [5], the European Space Agency [6] and the Joint Electron Device Engineering Council [7]. Method 1080 (single-event burnout and single-event gate rupture test) in standard MIL-STD-750F [8] is the test method for power metal–oxide semiconductor field effect transistors (MOSFETs) in heavy ion irradiating environments. The radiation resistance needs to be justified by type testing or technical analysis during the acceptance process for commercial items as equipment important to safety. However, most of the qualification programmes for commercial grade smart devices do not consider radiation as an environmental stressor.

2.2.8. Potential for common cause failure

A significant issue raised by the use of programmable devices of all types, including smart devices, is the potential for CCF, which depends particularly on the specific architectural solution of all the systems in a plant that may contain smart devices. Most of the earlier work on smart devices, such as standard IEC 62671 [3], focused on the qualification of individual smart devices or families of such devices, with only very limited consideration of CCF across multiple layers of protection.

The subject of CCF in general is comprehensively covered in several existing publications that consider the equipment and system containing programmable elements. Valuable references include NUREG/CR-6303 [9], United States Nuclear Regulatory Commission (NRC) Branch Technical Position 7-19 [10], VDI/VDE 3527 [11] and IEC 62340 [12]. Ongoing standardization projects that address CCF and its mitigation are IEC 60880 [13], IEEE 603 [14], IEEE 7-4.3.2 [15] and IEC 61508 [1].

The cited guidance documents (aside from IEC 62671 [3]) are typically focused on the system level. However, it is now understood that the component level aspects of CCFs have to be addressed during qualification. Information on CCF aspects specific to smart device integration into the I&C architecture of the full plant is provided in Section 3.

2.2.9. New failure modes

Often, smart devices come with new and complex possible failure modes, usually related to software. Hardware failure modes are subject to detailed analysis by FMEA and follow-up incorporation of means of automatic detection of the dominant dangerous failure modes. These measures are required by IEC 61508 [1]. The internal structures of smart devices are usually not disclosed, and the internal structure and interconnectivities may be more complex than necessary for the primary function, since commercial and industrial devices are usually designed to maximize market potential. An assessment of a smart device for use in an NPP system requires analysing failure modes caused by unexpected behaviour of the smart device and evaluating the impacts on plant safety. Many unexpected behaviours — such as spurious actuation, incomplete activation, shifted timing of signals, electron migration, voltage fluctuation and signal routing — could be possible, so very thorough V&V may be required.

2.2.10. Possibility of counterfeit items

Non-conforming, counterfeit, fraudulent and suspect items (NCFSIs) found in today's global supply chain present challenges to the quality control of all equipment intended for use in safety applications in NPPs; however, the concern is more acute for smart devices. Counterfeit devices are of lower cost and quality, and so may contain lower grade components, from capacitors to chips, and less well defined software, all of which is difficult to detect. NCFSIs in smart devices can result in non-compliance with regulatory requirements. They can also prevent a smart device from performing its intended safety function or cause other safety components to fail to perform their intended safety functions.

Effective procurement programmes as part of quality assurance (QA) can help with the prevention and detection of NCFSIs and reduce the likelihood of their introduction into the supply chain for smart devices. The characteristics of an effective procurement programme include the involvement of engineering staff in the procurement and product acceptance process; effective source inspection and, in some cases, third party supplier audits; receipt inspection and testing programmes; and thorough engineering based programmes for review, testing and justification of commercial grade smart devices for use in safety applications.

Vendors, suppliers and owners of NPPs all need to verify the genuineness of smart devices destined for safety applications. This verification includes extensive inspections of smart devices' critical physical characteristics, combined with rigorous performance testing, to provide reasonable assurance that smart devices will perform their intended safety functions.

2.2.11. Requirements for strict version control and material source control

With the rapid advancement of digital technologies, vendors and manufacturers of smart devices upgrade their products often, especially in the case of industrial or commercial grade devices. Upgrades may include introducing changes to subcomponents, applying new firmware or software versions or using new manufacturing processes. This means that vendors and end users need effective configuration management programmes to properly identify all modifications to smart devices, and strict version and material source control for smart devices has become increasingly important.

An adequate QA programme for applications in systems important to safety can help with version and material source control for smart devices, especially if it has the following QA elements:

— Design control;
— Procurement document control;
— Control of purchased material, equipment and services;
— Identification and control of material, parts and components;
— Disposition of non-conforming materials, parts and components;
— Review of corrective actions and programme effectiveness.

All these elements can provide good control of the version and material sources for smart devices.

2.2.12. Potential for hidden smart devices within otherwise conventional devices

Industrial manufacturers are increasingly faced with technological obsolescence, so they may be forced to seek new subsuppliers to obtain form–fit–function replacement components for their devices when these are required. These third party suppliers may substitute a smart device within the form–fit envelope and thereby expose the nuclear end user to an unknown functional deficiency. This is an additional motivation for NPPs to require detailed traceability of components during procurement and possibly some design review of replacement devices.

2.2.13. Interface with other technologies within the target system

Smart devices are normally implemented within other target systems that can use slightly or significantly different technologies. This situation can potentially lead to compatibility issues if not properly treated. Some compatibility issues may be hidden, especially if the technologies are not so different, therefore careful assessment of smart device boundaries within target systems is important to identifying and addressing such issues. The aim of assessment is to acquire enough evidence for demonstration of all relevant behaviour properties of the smart device, the target system and their boundaries.

2.2.14. Sensitivity to the quality of existing power supplies

Microprocessors and HPDs, such as field programmable gate arrays (FPGAs), utilize a number of internal low voltage power supplies derived from 24 V DC or AC power sources. Whenever the source power is interrupted completely for a duration of seconds or more, these devices usually restart automatically, and their functionality for such power fail–restart processes is a normal part of the design for industrial devices.

Problems may occur, however, if the source power interruption is shorter than approximately 1 s (possibly for only a few cycles of an AC source), which can lead to the different voltages needed by the chip behaving differently during the source voltage drop and subsequent recovery. Instances have been observed where the chip has not rebooted as intended. This creates a failure mode that does not have a counterpart in analogue devices.

While the logic components of a smart device (i.e. microprocessors or HPDs) probably draw less current in steady state operation, this may not be true during power-up. If the smart device includes any actuating device, such as a valve positioner, it will undoubtedly draw essentially the same current as the device that it replaces, but such components may be activated during a power-up to compensate for actuation as power fails. Thus, it is possible that inrush current loads during a power restoration may exceed expectations based on a previous analogue system. In the worst case, this may require a power-up sequencing unit to turn on each device sequentially, starting with the device with the largest inrush current.

In general, the characteristics of the power requirements of a smart device (in terms of the quality of the supplied power) and the quality of the existing power supply need to be reviewed to ensure that they are compatible.

2.2.15. Sensitivity to cabinet temperature

Elevated temperatures have adverse impacts on semiconductors in general and therefore on smart devices. Semiconductors age faster at higher temperatures, which inherently means that their failure rate is increased, and if several smart devices are located in the same cabinet, there is the potential for CCF.

While individual devices are usually adequately cooled by their own cooling fans, overcrowding in a cabinet, warm inlet air or failure to provide an active cooling system may lead to elevated temperatures. Appropriate measures — such as enhancing the air ventilation system, adding an exhaust fan system or design change in a cabinet — may be needed to keep temperatures in a suitable range for smart devices. The methods to predict cabinet temperature are the same as for non-smart devices.

2.2.16. Hardware qualification

Smart devices need to undergo the same environmental qualification as their analogue counterparts, but with the current technologies using extremely small sizes of gates (of the order of a few tens of nanometres), smart devices are likely to be more sensitive to radiation than corresponding analogue devices. Gamma radiation can interact with the packaging of a chip and generate alpha rays, which may corrupt the configuration or the memory on a chip. Several technologies are used by chip manufacturers, some of which are more resistant to such damage or are capable of auto-correction within a short time. The sensitivity of smart devices may be a concern, depending on the environment to which they could be exposed.

2.3. LICENSING TOPICS RELATED TO QUALIFICATION

This section discusses current licensing challenges faced by both regulators and end users that affect the preparation and acceptance of justifications or approval for smart devices to be used for safety applications in NPPs.

2.3.1. Review and selection of an approach based on recognized practices

The initial task in planning assessments of smart devices is to establish the process. There is not a single, universally accepted approach to assessing a smart device's suitability. A Member State seeking to establish such an approach may base it on IEC 62671 [3] or another standard that addresses the same scope, namely assessing a device and its certification to a suitable standard. It is also possible for a Member State to accept another set of standards as equivalent to the IEC standard for this purpose. One example is Canadian Standards Association (CSA) standard N290.14-15 [16].

Another possibility is the approach adopted in the United States of America, which uses the Electric Power Research Institute (EPRI) Technical Reports TR-3002002982 [17] and TR-106439 [18], which have been reviewed and endorsed by US NRC for justification or dedication of commercial grade items and services used for NPPs. EPRI TR-107330 [19] can also be used as a reference, although it was created for the qualification of commercially available programmed logic controllers. EPRI TR-3002002982 [17], which is specifically endorsed in Regulatory Guide 1.164 [20], with exceptions and clarifications, is the updated commercial grade item dedication guidance to supersede EPRI NP-5652 [21] and TR-102260 [22].

The approach used in France is provided in the RCC-E standard [23], which provides two qualification paths for smart devices: either using IEC 62671 [3] or using smart devices that are already certified to IEC 61508 [1] and for which appropriate evidence is available. A graded audit by the level of safety integrity level (SIL) requirement is also determined by the safety class.

The approach in the United Kingdom (UK) is derived from safety assessment principle ESS.27 [24], which requires the assessment of the production excellence of a smart device (including the quality of the development process and of its V&V) and additional independent confidence building measures to confirm the suitability of a digital device for a nuclear application. Additional details on the regulatory expectations in the UK and on the tools typically used for the qualification of smart devices are provided in Annex IV. Information on applicable practices in Member States for smart device selection and review practices are provided in Annexes I and IV.

2.3.2. Capabilities of organizations charged with qualification

Organizations supporting qualification need to be able to demonstrate an appropriate QA programme and the competency of their staff.

2.3.3. Limited access to detailed design information

Vendors for smart devices are usually reluctant to reveal the design details of expensively developed solutions to difficult problems, which are considered to be valuable intellectual property and carefully guarded. In particular, the embedded software in smart devices is usually not available or accessible to either the end user or the regulator. However, vendors or suppliers need to provide an adequate level of design, implementation, manufacturing and testing process information for licensees and their regulators to review the development process as part of their justification. Hence, a process is needed to qualify off the shelf industrial or commercial smart devices on the basis of the evaluation of both their development process and their specific functional properties, rather than expecting qualification by virtue of development according to a nuclear standard and specific functional properties.

2.3.4. Inconsistencies in structure and intent in the quality assurance programme

The search for applicable smart devices may lead to selecting devices from multiple countries that have different national QA approaches for nuclear applications. This may require mapping different QA systems to the system adopted by the regulator in the target country. This issue is not unique to smart devices, but the presence of software is a complication that must be managed.

For example, the structure of the CSA's CAN3-Z299 series of standards [25] encompasses multiple levels of quality, ranging from commercial (at Level 4) to the equivalent of Class 1E (Level 1), and the structure at each level is quite different, requiring a cross-reference matrix to determine the degree of commonality. As another example, the QA programme requirements in the United States of America for the design and construction of NPPs are included in Ref. [26]. The American Society of Mechanical Engineers (ASME) Nuclear Quality Assurance-1 (NQA-1) standard [27], codified in Ref. [28], describes the code requirements and implementation of all QA criteria in more detail for end users. Reference [29] — which specifically endorses, with certain clarifications and regulatory positions, the requirements in Refs [27, 30, 31] — describes methods considered acceptable for the regulatory requirements in Ref. [26] for establishing and implementing a QA programme for the design and construction of NPPs. Thus, the inconsistent structure and intent among different QA standards can pose a challenge for a state when setting up its own QA regulation for the licensing process of smart devices.

2.3.5. Differences in design criteria for software with high safety significance among Member States

Software for safety functions in other industries is usually created to comply with standards such as IEC 61508 (Vol. 3) [1], which requires a guaranteed maximum response time for SIL 2 and higher. In Member States applying the IEC nuclear standards, software for Category A functions must be designed to meet criteria such as determinism, as required in IEC 60880 [13]. Some nuclear standards that are applied in specific Member States, such as RCC-E [23], relax this constraint somewhat and require only predictability in the scheduling of software functions, which corresponds to the IEC 61508 [1] criterion. This means that there may be limitations in the highest class to which a smart device might be qualified in some States.

2.3.6. Differences in software verification and validation requirements

For systems and equipment used for safety functions, V&V is required to be performed according to accepted international standards, such as IEEE 1012 [32] and IEC 60880 [13], and independence is usually required for the verification as each phase of the life cycle is completed. Smart devices being qualified for use in systems important to safety can be verified and validated to IEC 61508 [1], which has a slightly different emphasis on what testing is required and when independence is formally required. Under IEC 61508 [1], the mandatory requirement for independence applies during assessment by a certifying body, at which time there is a complete review of all work, including the V&V, from beginning to end. The degree of independence and expertise of the assessor is dependent on the SIL capacity of the device and other factors.

In addition, independent V&V, required in the above standards, calls for three different types of independence: organizational, technical and managerial. The independence aspect of the V&V is vital, yet it could pose another challenge for the smart device's vendors.

2.3.7. Configuration management requirements

Maintaining good configuration management and control, as required in some recognized standards (see, for example, standards IEEE 828 [33] and IEC 61508 [1]), plays an important role in the life cycle of smart devices. The minimum set of software activities for configuration management required in the standards usually includes identification and control of all software designs and implementation, functional data (including parameter values), design interface, control of software design changes, and software documentation for users, operating and maintenance staff. Provision for authenticating the version of smart device firmware or software is essential. Additionally, the smart device needs to provide a means to lock the user configuration data.

2.3.8. Differences in expectations of suitability analysis for different applications

After a commercial or industrial grade smart device is qualified for a target application, it may be assumed to be suitable for any safety application. This might not always be appropriate, because of the limitations of existing qualifications on some aspects, such as the qualification scope, application specific environments and functional requirements. The qualification process can become challenging, especially for some Member States where qualification efforts carried out might not be commensurate with the safety classification or significance of the applications.

3. CONSIDERATIONS FOR COPING WITH COMMON CAUSE FAILURES OF SMART DEVICES

3.1. COMMON CAUSE FAILURE CONSIDERATIONS WHEN USING MULTIPLE SMART DEVICES IN INSTRUMENTATION AND CONTROL ARCHITECTURES

This section provides an overview of the considerations for CCF when using multiple smart devices in a plant architecture. These considerations are equally valid for all components in which the logic is implemented using software or firmware. The main implementation of these devices is in I&C systems, so this section focuses on these systems. However, it is important to recognize that smart devices are increasingly used in mechanical systems (e.g. smart valves) and electrical systems (e.g. smart protective relays), and the same considerations for CCF are equally applicable to these systems.

As discussed in Section 2, in modern plant design and plant upgrades, smart devices tend to be preferred to traditional analogue components for a number of reasons (e.g. obsolescence of analogue components, availability on the market, cost). Since smart devices can be installed at different levels in the plant architecture (possibly delivering functions at different safety categories), their deployment introduces new challenges in that the failure associated with software or firmware introduces the potential for CCF.

For NPPs, protection against accidents can be provided by plant architecture designs through multiple layers of protection, each of which is intended to be independent, particularly during accident conditions. Smart devices installed in more than one layer of protection might fail during the same fault sequence as a result of a CCF, and this failure mode may be difficult to determine. However, the defence in depth concept is expected to provide layer to layer independence, and for operational reasons, it is common to use identical smart devices in multiple channels of a system in any one layer of protection. Therefore, the use of smart devices in more than one layer of defence typically raises CCF concerns that need to be considered by the NPP operator.

Although CCF is not unique to smart devices, the presence of software or firmware in smart components means that they present additional challenges compared with their analogue counterparts, as a result of the introduction of additional sources of systematic failure. Unlike random failures, a systematic failure is "failure, related in a deterministic way to a certain cause, which can only be eliminated by a modification of the design or of the manufacturing process, operational procedures, documentation or other relevant factors" (see IEC 61508 (Vol. 4) [1]). Such failures are potentially dangerous from a CCF perspective, as multiple devices can be affected. Examples of causes of systematic failures in smart devices include the following:

— Erroneous or incomplete specification of requirements;
— Undetected errors introduced during software or firmware development;
— Errors in configuration control, leading to incorrect equipment settings;
— Vulnerability to cyberthreats[4];
— Vulnerability to other external factors, such as stressing environmental conditions (e.g. the effects of electromagnetic interference (EMI), radiofrequency interference (RFI) or radiation on digital components);
— Inclusion of common low level software components that may not all be revealed in the documentation available to the user (or even the manufacturer, where reliance is placed on lower levels of the supply chain).

Systematic failures can arise from the software or firmware of a smart device as a whole (e.g. the same product is used in two or more places within a plant architecture design) or from common software or firmware elements used in various smart devices of different types (e.g. operating systems, tools or embedded modules that are not disclosed). Annex I provides additional details on how smart devices of different types can share common elements (e.g. common software) and how diversity can be used as a means to reduce CCF risk.

The failure of multiple smart devices deployed in the plant architecture design requires a mechanism for triggering systematic failures to cause the devices to fail simultaneously (or in a limited time period). The triggering mechanisms are associated with factors that may be beyond the design of individual smart devices and arise from the context in which they are used (e.g. sharing input data, maintenance or other operational issues). Annex II provides additional insights on triggering mechanisms and how these can lead to the failure of multiple smart devices.

3.2. ASSESSING COMMON CAUSE FAILURES CAUSED BY SMART DEVICES IN THE PLANT ARCHITECTURE

Because of the complexity of the digital systems associated with software and the inherent limitations in testability, there is a need to address the design considerations that are applicable when introducing smart devices or other

---

[4] Cyber vulnerability may compromise multiple devices in the plant architecture and have a significant impact from a nuclear safety perspective, depending on the design of common devices. See Annex II for additional insights.

digital devices into a plant architecture to avoid CCF. Since it is impracticable to demonstrate that there are no residual faults in the design of smart devices, the risk of CCF has to be addressed at the plant architecture design level and by smart device selection (e.g. by implementing diversity). Annex II provides examples of how CCFs among multiple smart devices can affect the overall I&C architecture. Annex II provides further examples of how CCFs among multiple smart devices can affect the overall architecture of the plant electrical system.

Figure 1 outlines an iterative process that can be applied to plant design, device selection and CCF analysis. The first part of Fig. 1 (top left) highlights the process of identification of the smart components and selection of candidate devices, typically based on the components' compliance with the requirements of the proposed application. Examples of typical safety and operational requirements include reliability targets, independence and diversity constraints, accuracy, operational range and response time. Any non-conformance with this set of requirements needs to be resolved before proceeding to the next steps shown in Fig. 1.

The second part of the process shown in Fig. 1 (centre and right) identifies the potential need for iterations between CCF and design before convergence on

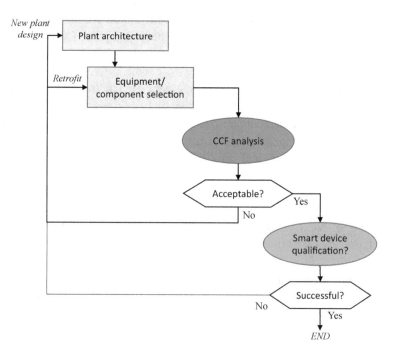

FIG. 1. Example of iterative process applied to the overall plant architecture design and CCF analysis.

an acceptable solution. For retrofits, the intention is not to redesign the plant but rather to ensure consideration of the CCF implications of the deployment of a smart device type in the existing plant architecture. For retrofits, if component reselection is not successful, because of CCF concerns or other issues, then it may be necessary to reconsider the scope of the modification to ensure that plant architectural design is resilient to CCF and other issues. For new plants, the design process can be optimized, as there may be architectural solutions that could avoid the impact of potential systematic failures affecting multiple systems or safety loops.[5] Examples of actions to improve resilience against CCF include the following [34, 35]:

(a) Adopting fail-safe solutions at the plant design level (e.g. mechanical actuators not relying on I&C activation).
(b) Modifying the plant architecture to increase functional diversity (e.g. detecting different physical parameters, or the same parameter at different stages of a fault sequence).
(c) Introducing equipment diversity at the I&C level, for example in the following ways:
    (i) By using both smart and non-smart devices (e.g. smart for complex functions and non-smart for simpler safety functions);
    (ii) By using different smart devices where it is possible to demonstrate to a high level of confidence that there are no common modules (disclosed, embedded or other) contained within the two devices.

Different levels of resilience against CCF can be achieved by using the approaches above, and these could be applied proportionally, depending on the categorization of safety functions. For example, using a combination of smart and analogue devices is typically more resilient than using different smart device models from the same vendor. However, the use of diverse devices (smart, analogue or a combination of the two) can have an impact on the complexity of the architecture, operations and maintenance (e.g. in terms of stocking of spares, maintenance, operational instructions, surveillance testing, human factors and training). Some of these aspects are particularly relevant from a nuclear safety perspective and need to be managed appropriately to avoid increasing the operational risk (e.g. potential for maintenance induced errors caused by using different smart device models).

---

[5] Architectural solutions are a powerful way of reducing overall plant risks and can also help to reduce the qualification effort of a smart device, for example by reducing its classification requirement or the compensation of gaps identified in the qualification process (see Section 4).

Once a candidate device has been selected, a level of CCF analysis is needed if the device is to be implemented in multiple applications in an NPP. This activity is distinct from the smart device qualification (see Section 4), which focuses on the use of a smart device at a certain safety class. The importance of this verification is that multiple failures are typically more significant from a nuclear safety perspective than are individual failures. The first step in the CCF analysis illustrated in Fig. 1 is to understand the failure modes of individual smart devices (including spurious actuation). The focus of the CCF analysis is to determine the impact of each failure mode when it affects multiple devices and the consequences for the overall plant architecture. CCF analysis needs to involve both engineering competences (e.g. expertise in the failure modes of smart devices, as well as in-depth understanding of the applications in the plant) and fault study expertise (e.g. understanding how the CCF can affect the plant and whether the scenario is covered by the existing safety analyses). The extent of the CCF analysis may also depend on the safety category of its applications in the plant architecture. Annex II provides additional insights into the elements to consider in CCF analysis. The results of the CCF analysis depicted in Fig. 1 need to be documented in order to demonstrate the design options considered and the rationale behind the solution selected.

After convergence has been established between the architecture design and the CCF analysis, the smart device qualification step (orange colour in Fig. 1) needs to be completed. The principles to be addressed as part of the smart device qualification are outlined in Section 4.

3.3. EXAMPLES OF ARCHITECTURAL SOLUTIONS TO COMMON CAUSE FAILURES

Smart device selection is often driven by conservative decision making at the overall plant and I&C architectural design level. A typical solution implemented in many NPPs features the following:

— The adoption of device diversity in different layers of defence in depth;
— The use of the same smart device in different redundancies (or channels).

The rationale for the former is embedded in the concept of defence in depth and the requirement for independence between different layers, which in practical terms means making the choice of using different devices to minimize the potential for CCF. This is often implemented with additional requirements for signal segregation and maintenance of diversity requirements among smart devices belonging to different levels of defence in depth. The selection of the same smart device for redundant configurations may be supported by deterministic analyses that show that CCFs of smart devices in a system are bounded by the single failure criterion, requiring the plant to be resilient to the loss of a whole system.

New NPP designs may implement innovative features (e.g. passive systems, fail-safe features) or different safety targets (e.g. lower core damage frequency), which may require alternative architectural design considerations. For example, in some designs, implementing additional diversity within a layer of defence in depth may be acceptable in lieu of requiring multiple independent systems. In others, the use of diversity within each layer of defence in depth may be necessary to achieve more stringent safety targets (e.g. the single failure criterion). Similarly, for modifications at operating NPPs or legacy facilities, the architectural constraints or safety requirements may be such that different architectural solutions may be needed to address device obsolescence. In these cases, the process outlined in Fig. 1 can provide a framework for the justification of the adequacy of the overall I&C architecture and the smart device selection.

It is recognized that the detailed approach adopted to address the risk of CCF when using multiple smart devices will depend on the specific architectural design of the NPP.

3.4. COMPUTER SECURITY CONSIDERATIONS ON THE USE OF SMART DEVICES

Because of their digital nature, smart devices can be susceptible to cyberthreats. Smart devices offer a level of resilience against cyberthreats compared with other complex digital systems, as follows:

— There is limited potential to reprogram smart devices (although they can be reconfigured) compared with other digital systems (e.g. programmable logic controllers and other I&C platforms).
— Changes in parameters or configuration generally need physical access to the smart devices, which can more easily be protected compared with networked systems[6].

---

[6] In some architectural designs, smart devices can be networked together, which introduces additional cyber vulnerabilities to be considered. Examples in electrical power systems are discussed in Annex II.

There are, however, various scenarios in which cyber vulnerabilities of smart devices can be exploited, including the following:

— Access to backdoors in smart device software (e.g. for maintenance teams);
— Installation of a counterfeit device providing remote access to smart devices (see also Section 2.2.10);
— Hacking of smart device manufacturers and introduction of malware into their devices;
— Implementation of hidden malicious code in the libraries or tools used for the smart device development;
— Use of confidential information, such as administrator passwords, in a series of device types, resulting in multiple devices being vulnerable if the password is retrieved.

Some of the cyberthreats identified above arise from the supply chain. For example, malware inserted into a software library or module could affect several devices. Similarly, malware inserted into a software tool such as a compiler could affect all devices that are produced using that software tool.

The key challenge introduced by cyberthreats compared with other systematic failures, such as unintended software flaws or errors in requirement specification, is that cyberthreats typically change rapidly and can be designed to simultaneously target multiple smart devices, increasing the risk of a CCF. In general, because of the rapid changes of cyberthreats and challenges in the supply chain control for a commercial component, it is difficult to assess the adequacy of the protection against cyberthreats and to predict their consequences. A conservative approach is therefore to assume that an attack may happen during the deployment of a smart device type. CCF analysis can be used to determine whether the impact of such an attack would be acceptable — for example, whether a vulnerability in a smart device type could be exploited to compromise different security zones and what the consequences would be for nuclear safety and security. Overall architectural solutions are generally an effective means of reducing the impact of any cyberattack, as is the use of particular technology types whose vulnerability to cyberattack may be easier to identify and manage (e.g. use of HPDs, such as FPGAs, instead of microprocessors). Additional considerations on CCFs resulting from computer security vulnerabilities are discussed in IAEA Nuclear Security Series Nos 17 (Rev. 1), Computer Security at Nuclear Facilities [36], and 33-T, Computer Security of Instrumentation and Control Systems at Nuclear Facilities [37].

4. SMART DEVICE QUALIFICATION

4.1. OVERVIEW

The focus of this section is on the generic qualification of a smart device for a nuclear application. Before the practical implementation of a qualified device in a nuclear application, additional steps need to be completed to ensure that the application falls within the boundary of the generic qualification (see Section 5.3.2 on the suitability assessment). In order to efficiently complete these steps, it is important that the qualification be supported by a clear restriction of use (see Section 4.1.5 on restrictions of use).

The qualification process examines a candidate smart device for its suitability for use in an NPP. This process requires examining a number of criteria, including the following:

— Functional and performance suitability for the application, including functionality, robustness and reliability;
— Evidence of the adequacy of the design and its implementation and documentation;
— Criteria and constraints for integrating the device into the target plant architecture;
— Criteria for preserving the qualification over time.

The intent is to identify smart devices that are functionally and otherwise suitable for an application in an NPP. Although prior certification of a device to a non-nuclear standard, such as IEC 61508 [1], can offer some advantages in the qualification process for a nuclear application (e.g. evidence is already available), this does not represent an alternative to the qualification process itself. The qualification has to cover both software and hardware and may be generic for a range of applications or specific to one application.

The qualification process typically consists of defining the prerequisites and objectives, planning the qualification activities, performing the qualification analysis and reporting the results.

4.1.1. Prior to commencing the qualification of a device

Qualification is intended to ensure that a successfully reviewed device meets the clearly identified requirements for its specific applications. To ensure this, the following processes need to be established:

(a) Ensuring that the requirements imposed by the application in the NPP are clearly and completely defined and then translated into a complete specification for candidate devices. This specification will address the functionality required of the device, its performance requirements, environmental requirements, safety classification and reliability requirements with respect to both systematic software failures and random hardware failures.

(b) Preselecting available devices for apparent compliance with the plant. Factors to consider in the selection of devices are the following:
   (i) Availability of manufacturers to support the qualification process and share key information needed for this purpose;
   (ii) Operating track record of similar process applications in high integrity industry and alignment with the target system class;
   (iii) Availability of accredited certifications against relevant safety standards (e.g. IEC 61508 [1]);
   (iv) Suppliers' track record in producing high integrity applications.

(c) Determining the target envelope for the qualification for a single product, a family of products or several variants of the same product. This will later influence the qualification plan.

(d) Determining whether an existing qualification already covers this device, and if so, whether this can be reused or referenced (see Section 4.1.4).

(e) Determining whether there are already other smart devices installed or planned for installation in the NPP. If so, then consideration of CCF will become important, particularly where Category A functions are involved.

Once these points have been addressed, and thus the scope of the qualification identified, the next step is the qualification itself.

4.1.2. Qualification objectives

The prerequisites lead to several objectives, including the following:

(a) Definition of the scope of the qualification; for example, by addressing the following aspects:

  (i) Whether a specific product or a larger family of products is to be qualified;

  (ii) Whether only one plant application or a range of applications is to be considered.

(b) Confirmation that the following requirements arising from the plant application(s) are met:

  (i) Capability to meet the functional and performance requirements;

  (ii) Hardware reliability and environmental qualification;

  (iii) Required safe failure modes and failure rates (both dangerous and spurious);

  (iv) Maintainability and testability constraints (e.g. whether it is necessary for maintenance or testing to be possible without shutting down the plant);

  (v) Human factors related to the operation and maintenance of the device (including consistency with existing equipment);

  (vi) Constraints on device selection based on plant architecture, including constraints based on the possibility of CCF with other layers of protection;

  (vii) Required device lifetime and duration of device support from the manufacturer.

(c) Confirmation that the candidate product is otherwise suitable for application in systems important to safety, including the following aspects:

  (i) Adequate quality of product development, as the prime indicator of freedom from systematic faults;

  (ii) Operational experience with the device;

  (iii) Adequacy of the documentation, both for review of the design and for all operational needs;

  (iv) Adequate attention to any additional considerations, such as security requirements.

(d) Reporting of the qualification results, as follows:

  (i) Identification of the limits of use under the qualification (e.g. safety class, environment);

  (ii) Identification of precautions required, such as to manage non-essential functionality;

  (iii) Potential CCF triggers (e.g. software library used) that could affect other smart device qualifications.

A number of the issues listed above are discussed in more detail in Section 4.2, and the entire scope is addressed in the qualification plan (see Section 4.4.1).

4.1.3. Software and hardware qualification

Qualification of smart devices typically consists of software and hardware qualification, with the objective of achieving adequate confidence in the dependability of the device to correctly perform its safety functions in both operational states and accident conditions.

Hardware qualification for smart devices involves the same requirements as for analogue devices, but there are a few issues unique to smart devices that need to be considered during the hardware qualification. Some of these are specified in Section 4.2.4.

Software qualification is made necessary by the complexity of software, whether the device contains software or is considered to be hardware that was designed using software based tools. Software introduces multiple types of failure that do not exist with analogue devices, such as the following:

— A change in compiler version, or in the options used with a compiler, may result in a different assembler code, thus introducing an unexpected change in the behaviour of the software.
— A flood of external events related to a plant upset may cause a stack to overflow, leading to unpredictable results (e.g. because code or data are overwritten by unexpected data).
— A chip may be impacted by radiation that can change the state of memory and the execution path of software using that area of memory.
— Often designers incorporate software modules from other sources, such as manufacturers' libraries, and these may contain hidden errors or features and may be common to more than one device in the plant (thus creating the potential for CCFs).
— Testing may be incomplete, and some combination of inputs during a transient may lead to the execution of untested code. Software is extremely difficult, if not impossible, to test fully, particularly if it is not specifically designed to be extremely simple (such as a simple loop with limited branches and without interrupts).

4.1.4. Generic qualification versus specific qualification

For both software and hardware, qualification may be generic or specific. Generic qualification is based on selecting a suitably high level of qualification that would be intended to envelop all possible applications of the smart device. This can minimize the work needed to justify the applicability of the qualification for many specific applications, but it is essential that the scope of use of the device be understood and correctly specified. Generic certification usually provides evidence of the correctness of the design, the suitability for the class of application and the criteria for preserving the qualification.

Specific qualification considers all aspects of a qualification and provides a justification for the use of a smart device in one specific application. A specific qualification can reference a pre-existing generic qualification to the extent that the generic qualification encompasses the specific application and hence may address only the issues specific to a given application. Thus, a generic qualification of a pressure transmitter installed in a reactor trip system may apply for a specific generation of NPPs, but not for use in an engineered safety features actuation system if the environmental conditions are different.

Examples of aspects requiring specific consideration include the following:

— Suitability of power and signal interfaces with other devices in the system;
— Possibility of CCFs with other devices already installed, or due to be installed, in the plant;
— Hardware qualification details based on where in the plant a device will be installed and the seismic or environmental stresses to which it will be exposed.

Examples of aspects likely to be covered by either a generic qualification or a specific qualification include the following:

— Quality of the development environment, including the software development process and supply chain management;
— Analysis of the effects of secondary functions;
— Detection of undocumented features;
— New modes of failure compared with a device being replaced;
— Extent of user configurability and locking mechanisms;
— Counterfeits;
— Hardware qualification;
— Expected lifetime of the product (obsolescence).

4.1.5. Restrictions of use

On completion of the qualification, the allowed application(s) of the device, plus any restrictions of use of the device, need to be defined and recorded. Examples include the following:

— Properties or attributes of the smart devices not of interest for the target application, hence scoped out from the qualification (e.g. propensity for spurious actuation, low sampling frequency of inputs that could require filtering of input signals with high frequency noise);
— Configurations of the device not included in the qualification (e.g. use of a highway addressable remote transducer or fieldbus, use of an external human–machine interface);
— Assumptions adopted in the qualification (e.g. impact of proof testing interval or operating temperature on hardware reliability);
— Consideration (or not) of the robustness of the device to spurious actuations;
— Applications not recommended for use of the device, such as architectures where this or a very similar device is already in use, or where the device must function for an extended period without maintenance.

4.2. QUALIFICATION ATTRIBUTES AND CRITERIA

This section highlights a number of the primary criteria to be considered during qualification, describes the technical issues and identifies ways to verify the suitability of a candidate device for application in systems important to safety in NPPs. Requirement 63 of IAEA Safety Standards Series No. SSR-2/1 (Rev. 1), Safety of Nuclear Power Plants: Design [38], provides requirements on establishing "appropriate standards and practices for the development and testing of computer hardware and software…to be implemented throughout the service life of the system, and in particular throughout the software development cycle". These requirements generally apply to the selection and qualification of smart devices.

4.2.1. Compliance with functional and performance requirements

The functional and performance requirements are defined by the safety functions needed in the NPP. Initial screening of candidate devices can be accomplished by examining device datasheets. Additional information is available in the form of user manuals. For devices precertified to IEC 61508 [1], all required details are available in the device's safety manual.

The first step in verifying the compliance of a device's functional and non-functional requirements is to examine the validation testing performed by the manufacturer. All aspects of the device's functional and performance specification have to be traceable to specific black box test cases that validate the device's behaviour in terms of functional and performance criteria. For example, tests need to verify that outputs are within defined timing and value tolerances over the complete range of input signals and power supply voltages. The completeness of these tests can be judged in part by reference to the testing requirements established in standards such as IEC 60880 [13] and IEC 61508 [1].

Additional tests may be required as part of the qualification, depending on the safety class and the nuclear regulatory requirements, to determine, for example, the existence of functionality that is not necessary, functionality that is not declared or possible gaps in the manufacturer's testing. For example, a smart sensor may include an optional function to perform some advanced filtering of the signal that is not used in the NPP safety application. In such a case, tests could be performed to provide confidence that this functionality cannot be activated by plant events such as a power supply upset or momentary upset in an input signal. Additional tests could also be applied to verify that the device resumes operation correctly following a power supply upset.

It is also essential to confirm that the primary safety function will execute on time under all circumstances. Depending on the safety class, this may necessitate a design that guarantees determinism, or at least predictability. The qualification needs to examine how this is accomplished (e.g. by using simple loops without interrupts and a hardware watchdog). It has to also examine the possibility of subtle hardware faults impacting software operation and establish whether a flood of incoming events or a non-safety function could interfere with the safety function.

4.2.2. Adequacy of the development process

Paragraph 6.37 in SSR-2/1 (Rev. 1) [38] states:

"For computer based equipment in safety systems or safety related systems:

(a) A high quality of, and best practices for, hardware and software shall be used, in accordance with the importance of the system to safety.
(b) The entire development process, including control, testing and commissioning of design changes, shall be systematically documented and shall be reviewable.

(c) An assessment of the equipment shall be undertaken by experts who are independent of the design team and the supplier team to provide assurance of its high reliability."

The quality of the development process is the most important factor in minimizing the likelihood of systematic faults in a smart device. Aspects such as QA, supply management, configuration management, the development life cycle, the manufacturing process, the thoroughness and independence of validation and verification, the competency and training of staff and the presence of an obsolescence strategy are all important in determining the rigour and adequacy of the device development process.

International and national standards address these issues to varying degrees. Examples of such standards include IEC 60880 [13], IEC 62138 [39], IEEE 1012 [32] and IEC 61508 [1].

Compliance might be assessed by reviewing manufacturer documentation and performing a clause by clause compliance analysis to a suitable national or international nuclear standard. However, this will depend on the standard applied during the original development and also requires access to development documentation that can sometimes be difficult to obtain.

Different Member States have developed varying approaches to dealing with this problem. Evidence of a manufacturer's compliance with a widely accepted industrial standard such as IEC 61508 [1] is usually accepted towards the demonstration of development quality. IEC 62671 [3] provides a framework for assessing predeveloped smart devices of limited functionality, which may or may not have been developed to nuclear or industrial standards.

Member States typically follow a standards based approach, with some having developed country specific proprietary methods. These may involve different options for qualification and combinations of strategies, including the use of certification and other techniques, such as independent testing and 'proven-in-use' arguments. Annex IV provides examples of practices in Member States.

Some specific criteria to examine include the following:

— Use of a design life cycle that involves stages of development such as functional requirements, architecture design of hardware and software, high level software design and low level software design;
— Independence of detailed verification at each stage of the life cycle;
— Effective use of configuration management tools to control all design products (e.g. specifications, test procedures, board design, code, software tools);
— Design change control, including impact analysis;

— Use of design constraints such as coding standards and verification tools;
— Use of design analysis techniques to confirm the correct structure of the code (e.g. static analysis) and to measure the code complexity;
— Use of design tools (e.g. compilers) justified by certification or testing or otherwise demonstrated to be acceptable;
— Use of verification techniques and tools to confirm the degree of completeness of testing (e.g. tools that verify that every path in the code has been tested).

4.2.3. Confirmation that the device has a suitably low random failure rate

All hardware is subject to random failures. These tend to have physical causes and usually develop over time as a result of factors such as radiation exposure, corrosion, thermal exposure and mechanical stress. Such faults are random in nature, but statistical information can be gathered to calculate the probability of their occurrence. One of the objectives of qualification is to make sure that the random failure probability (or rate) is within the acceptable range based on the device's role in the safety function of the plant, as defined by its safety class and plant safety analysis.

Examples to be reviewed in the device production life cycle that are important for reducing the random failure rate include the following:

— The supply chain management;
— The hardware design;
— The manufacturing QA.

Factors that impact hardware reliability include low quality or counterfeit components from suppliers, errors in hardware design (e.g. power and signal landlines being too close together on a printed circuit board, failure to connect pull-ups or pull-downs to all ports on a chip), manufacturing errors (e.g. out-of-tolerance drilling or soldering) and damage during transport and handling. Consistent use of QA in the supply chain and manufacture can support the judgement as to the acceptability of the same or similar models of the device, even if they are manufactured later. A qualification assessment looks for evidence that a suitable QA programme exists and is carried out effectively by the manufacturer.

FMEA can be used to predict failures and improve the design. It is used to identify device failure modes and their impact, as well as to calculate the random failure probability from established failure rate databases that factor in environmental conditions. This analysis identifies the significant failure modes that lead to dangerous or frequent device failures, and this information is used to modify the hardware design to include redundancy, use more reliable components or add protective components, as well as to introduce automatic diagnostics to reduce the dangerous failure rate of a device. The qualification process verifies that these diagnostics are tested by fault injection tests to either drive the device to the safe state (for the application) or provide alarms if this is sufficient.

4.2.4. Confirmation that the device will withstand all operating conditions

A smart device needs to continue to operate correctly during all normal operation and accident conditions so that it can perform its intended safety function. Additional specific qualification activities need to be conducted if the conditions used for generic qualification of the smart device do not encompass the application specific conditions.

Most concerns related to hardware qualification between analogue and smart devices are covered in the methods used for analogue devices. These are defined in the existing relevant standards, as follows:

(a) Environmental qualification is well covered by IAEA Safety Standards Series No. SSG-69, Equipment Qualification for Nuclear Installations [40], and IEC/IEEE 60780-323 [41], which address peak temperature, pressure, humidity, radiation on accident dose, power voltage and frequency variations and ageing.
(b) Seismic qualification is well covered by IAEA Nuclear Safety Series No. SSG-67, Seismic Design for Nuclear Installations [42], and IEC/IEEE 60980-344 [43], which define how to demonstrate that the smart device can meet its performance requirements during or after one safe shutdown earthquake event preceded by a number of operating basis earthquakes.
(c) EMI/RFI qualification methods are well covered by IEC 61000-3/4/6 [44] and US Department of Defense interface standard MIL-STD-461G [45].
(d) Qualification against power disturbances is covered by IEC 61000-4 [44] and IEEE C62.41.2 [46].

Nevertheless, there are some specific issues applicable to smart devices that must be considered in the qualification, such as the following:

— Separation of power lines and signal lines on printed circuit boards to minimize interference with internal data transmission from power transients;
— Derating of all components, which provides a margin of robustness in all target environments and internal loads;
— Specific defences against variations in supply voltages and temperature;

— Software settings or parameters that must be adjusted for different environments;
— Environmental testing that includes continuous operation of the software and monitors all components of the smart device.

4.2.5. Confirmation of the adequacy of the user documentation of the device

User documentation includes manuals that support safe installation, including guidance on set-up and configuration, operation, maintenance, failure diagnostics and replacement of a device, as well as the use of any built-in security features. The qualification process includes checking all manufacturer documentation for applicability, completeness, clarity, accuracy and currency, and identifying any gaps that must be filled by the manufacturer or licensee.

The user-friendliness of the device for maintenance is also important: for example, it is preferable that the device not require iterative adjustments (e.g. of the zero and span for calibration) and that the documentation clearly explain how every needed maintenance procedure is to be executed. One may also ask the manufacturer how the user documentation is verified. An example of rigorous verification of maintenance instructions is to simulate a fault and then ask a maintainer to follow the manual to diagnose the fault.

Checking completeness can be facilitated by using a set of questions to confirm aspects of the device that, as experience suggests, may be omitted from the datasheet. This could mean ascertaining whether the device contains a microprocessor or HDL device, or looking for characteristics that indicate the device's suitability for the application(s), such as whether there are security measures for the device configuration and whether optional or undesirable secondary functionality exists.

It is important to clearly highlight any caveats, warnings or restrictions on the use of the device, as well as any special features. Language and diagrams have to be unambiguous and user friendly, whether natural or technical terminology is used. IEC 62671 [3] lists useful points to consider when assessing a device's user documentation.

4.2.6. Use of operating experience of the device

Most devices are certified to IEC 61508 [1] on the basis of the development process, but some devices are certified on the basis of prior use or proven-in-use arguments (this approach is very rarely used for SIL 3). Depending on the certifying body, this type of certification may be quite rigorous in terms of the quantity and type of operating experience needed (e.g. more than 30 million hours of applicable operation for SIL 3), which may include operating conditions that are at least as challenging as those of the target application, and the expected stability of the particular version of the product. For nuclear qualification, it is important to clarify that this certification path is used only where there are weaknesses in the documentation of the development process.

Nevertheless, operating experience can provide diverse supporting evidence independent of the development process. Since credited operating experience includes only the precise device versions to be qualified (or previous versions demonstrated to have only justified differences), it can provide general information on the performance of a smart device and support other evidence for the stability of the product in practice. A low number of revisions together with few or no failures would imply a good design process, even if it was not well documented.

If there is sufficient creditable operating experience, it may be possible to confirm failure rate predictions on the basis of an FMEA. For the software part of the device, credit is usually limited to the specific version(s) targeted for qualification. Typically, the standards (e.g. IEC 61508 [1], IEC 62671 [3]) also apply other constraints on creditable operation, particularly if results are based on manufacturer return data.

Additionally, for software that is actuated only under specific conditions, the prime consideration is the number of actuations, not the number of hours waiting for the conditions, as discussed in annex D to IEC 61508, Part 7 [1]. For these reasons, the operating experience of the device is primarily used as complementary evidence of the stability of the device design (which reflects the development discipline).

4.2.7. Confirmation of the device's resistance to cyberthreats

Cyberthreats to NPPs have two potential sources: malware deliberately introduced at the manufacturing stage and attacks after device installation in the plant. Although smart devices are not reprogrammable, they can typically be reconfigured, and protection is expected to be in place to prevent not just erroneous reconfiguration (e.g. during periodic maintenance or calibration) but also malicious activities (e.g. intentional security violations) from compromising a device's ability to deliver the required safety function. Access control is also recommended, although this is not a property of the device itself. The qualification needs to assess whether the following cybersecurity aspects have been considered:

(a) The application configuration is lockable to prevent changes. This means that parameters such as damping, signal range, limits and rate limits may be configurable, and therefore the configuration must be lockable and protected by a password or physical key.

(b) Device interfaces must be immune to cyberattack. The qualification review includes checking every input and output port (e.g. wired terminal or communications interface) for immunity to penetration by an external cyberattack. If the device includes features such as highway addressable remote transducers that are superimposed onto otherwise analogue signals, then it must be confirmed that the locking provisions can also prevent modifications via these features.

(c) Since it is possible for malicious intervention in the software of a smart device to occur at either the design or the manufacturing stage, the qualification needs to include a review of the cybersecurity measures taken by these parties to protect their own installations and those of their suppliers. Admittedly, designers and manufacturers are likely to be reluctant to reveal details of their protective regimes, but as a minimum the qualification rests on credible assurances that these regimes are in place, such as third party reports.

Guidance is provided in IAEA Nuclear Security Series No. 17 (Rev. 1) [36] and in IEC 62443 [47].

4.2.8. Review of factors that can impact a device's operation over time

Software, hardware and the application environment can all change the properties of smart devices over their operational life. Therefore, the qualification has to consider vulnerabilities in this regard. Changes in hardware characteristics, such as via instrument 'drift', are not specific to smart devices and need to be covered by other guidance documents, but semiconductor devices such as non-volatile random access memories are prone to lifetime limitations — in the number of read–write cycles, for example — and software can prematurely cause failures in such devices if not designed correctly.

Although software does not age in the way that hardware ages, all software contains internal data that are affected by changes over time, such as input signals read by the device and stored in a buffer, internal timers, interrupt states or flags. Internal state changes cause software behaviour changes, and over time this can, in some circumstances, lead to an error condition, such as software 'hanging'.

Other examples include memory leaks, buffer overflows, software reset caused by a date or time cut-off and data tearing[7] resulting from concurrency issues.

Furthermore, the state of the plant in which the device is supposed to operate may change over time, even if subtly. For example, many plants have undergone life extension and, in some cases, power uprating. Consideration needs to be given to how such changes would affect the device operation and whether the device configuration is appropriate to accommodate such changes.

Maintenance activities can introduce changes into the device via version upgrades or changes to configuration data, whether intentional or inadvertent. The connection and use of tools for diagnostics, maintenance and configuration changes has the potential to interfere with existing data stored on the device, and this needs to be considered.

The above potential for change in software behaviour over time needs to be analysed, risks need to be identified and methods for mitigating these need to be put in place. Such methods can be made part of a long term maintenance programme. Examples include device proof tests or periodic resets[8] to minimize the risk of undetected internal fault states or accidental changes to configuration data after maintenance activities.

4.2.9.   Confirmation of absence of any specific vulnerabilities

Each of the aspects of qualification addressed in the preceding subsections should be applied to check for vulnerabilities of the smart device that may be associated with the type of device, its technology or its intended use. Qualification seeks to provide confidence that the risk of residual errors from the vulnerabilities that might affect safe operation is sufficiently low.

Each type of underlying technology used to provide smart device functionality carries some vulnerabilities, and in many cases there are countermeasures supported by the manufacturer or guidance provided by standards to mitigate the effects of these vulnerabilities. Different technologies employ different levels of tool support, require different levels of complexity in code or the reuse of existing code, allow different levels of configurability and connectivity, and introduce different areas where security might be compromised. All these aspects may introduce the potential for faults, and this needs to be recognized and met with suitable scrutiny. Examples include the

---

[7] Data that span more than one unit of memory, such as two words, and can be retrieved by a single operation may be 'torn' if the task retrieving such data is interrupted after reading the first word and before reading the second word.

[8] To avoid maintainer induced common failures, maintenance activities on redundancies need to be staggered in time and executed by different people.

use of HDL based devices and the use of tools in creating FPGA based devices. Other more novel technologies, such as self-learning code and the use of artificial intelligence, could be subject to additional scrutiny (and are discouraged by IEC 61508 [1]).

The use of digital communication links allowing networking and data exchange between devices, as well as the complexity of supply chains, means that security requirements have to be formulated and considered. Guidance on the security of digital communication is provided in IAEA Nuclear Security Series No. 17 (Rev. 1) [36] and in IEC 62443 [47]. Licensees and regulators need to decide which technologies are acceptable for use in nuclear safety contexts and which standards are to be adhered to in the development and qualification of these technologies.

4.3. MANAGEMENT SYSTEM

Organizations responsible for the qualification of smart devices for nuclear installations generally develop, implement, assess and continuously improve a management system, in accordance with the requirements established in IAEA Safety Standards Series No. GSR Part 2, Leadership and Management for Safety [48], and in two supporting Safety Guides: IAEA Safety Standards Series Nos GS-G-3.1, Application of the Management System for Facilities and Activities [49], and GS-G-3.5, The Management System for Nuclear Installations [50].

The equipment qualification programme is generally derived from a QA programme that includes a variety of elements, such as equipment design control, procurement document control, manufacturing quality control, qualification assessment (e.g. testing, analysis, combined testing and analysis, experience), storage, installation and commissioning, installation surveillance and maintenance, periodic testing and documentation. Equipment qualification activities, including the assessment or reassessment of the status of qualified equipment, need to be performed in accordance with approved procedures and controls.

Traceability needs to be established between the qualification documentation, the conclusions from each qualification test or analysis and the configuration of the installed equipment, in order to ensure that the installed configuration corresponds to the qualified device configuration.

4.4. DOCUMENTATION

The qualification documentation of smart devices typically includes the following:

— The qualification plan;
— The qualification reports.

This section provides a skeleton of qualification documentation to identify the steps and outputs required to implement the qualification.

4.4.1. Qualification plan

Using either standard procedures or a device specific plan, the qualification plan is prepared and reviewed to establish the breadth and scope of the qualification and whether it is application specific or generic. The plan also identifies the target safety class of the smart device.

The plan is executed once it has been identified that the plant architecture requires a smart device to fulfil a function. The plan may address more than one candidate device. The qualification plan typically identifies the following:

(a) The scope and applicability of the qualification work, in terms of the following:
  (i)   The application or applications (safety functions) and the corresponding system class or classes (the depth of review is expected to be dependent on the integrity level required for the application of the smart device);
  (ii) How the selected qualified device will fit into the plant architecture and whether CCF is a concern;
  (iii) The nature of the qualification, namely whether the qualification is to be specific or generic (and to what degree).
(b) The candidate device or devices, including specific version levels, and the scope of the qualification analysis, whether software or hardware.
(c) The technical resources and the qualification needed to execute the evaluation work, such as suitably qualified and experienced people for the following tasks:
  (i)   To ensure a complete requirements specification, particularly in retrofit situations;
  (ii) To identify qualification criteria for the technology used in the device and study random and systematic sources of failure;

     (iii) To identify the safety class, potential CCF vulnerability with other layers of protection and the time domain qualification criteria in the target applications;

     (iv) To carry out an assessment of the product and its development life cycle against the specified qualification objective;

     (v) To carry out any additional activities, such as testing or analysis, that might be required during the course of the qualification.

(d) The tools and other resources needed to execute the evaluation work, such as the following:

     (i) Software tools to examine the susceptibility of the software to systematic faults;

     (ii) Specific test facilities to evaluate EMI/RFI qualification;

     (iii) Tools needed to evaluate operating experience.

(e) The standards and all the objectives to be considered in the qualification, and the contents of the qualification report.

(f) The methodology to apply, such as the following:

     (i) Clause by clause review of compliance to the standards selected;

     (ii) In the case of previously certified (to a non-nuclear standard) devices, performance of an audit of the certification (see Annex IV for details);

     (iii) A justification based on proven-in-use arguments according to the guidance in selected standards, with additional independent checks.

(g) The proposed criteria and means to establish the degree of confidence that the objectives have been met (e.g. definition of the level of review by the assessor of design documents, which may be dependent on the target class of the device).

(h) A framework for identifying and justifying the relative priorities of non-essential qualification objectives that may be difficult or impossible to meet using the available candidate devices, their manufacturers and the evidence available. An example of a non-essential objective would be the remaining support lifetime beyond the current remaining plant lifetime.

(i) Identification of qualification objectives that can be only partially met. Examples include non-compliance with clauses in a standard (e.g. formality of specifications or use of support tools) or weaknesses in documentation evidence to support standards compliance.

(j) The alternative compensatory measures to address the gaps identified in the qualification process (e.g. objectives are not met or only partially met). The alternative compensatory measures need to be justified. Depending on the nature and significance of the gap identified and the technology used in the device and the gap itself, it may be possible for it to be compensated by additional activities carried out by either the manufacturer, the licensee or a third party acting on behalf of the licensee. Such activities can sometimes be performed during a site visit, or they can be carried out as separate activities outside of the site visit. Depending on the extent of the gap and the feasibility of compensatory activities, the qualification of a smart device to a certain safety class may not be successful. Examples of compensatory activity types are available in IEC 62671 [3].

(k) Any special supplementary activities that could be carried out by the licensee to improve the confidence in the device's suitability. They may vary from commissioning tests and independent review of key documents (e.g. V&V test plan, FMEA) to more onerous activities such as additional dynamic testing, including statistical testing or source code analyses.

(l) The overall outcome of the qualification and recommendations for further steps, as described in Section 4.4.2.

(m) If the qualification is generic, how it will facilitate applicability judgements for specific applications.

At this stage, the licensee has to ensure the completion of appropriate review and verification by third parties involved in documenting the qualification. This includes the licensee's review as the informed customer of the qualification.

4.4.2. Qualification report

This section provides an example of the contents of a qualification report. This example allows for either a specific or a generic qualification approach. A generic qualification will need to be tailored additionally for the specific application.

A qualification report comprises the following:

(a) Introduction: scope and applicability of the qualification report. This section covers the points below and summarizes the results, while clearly specifying which sections of the report provide the detailed results:
   (i)   The intent of the qualification (i.e. the target systems and safety class);
   (ii)   The function provided by the devices covered in the qualification report (e.g. pressure transmitter, valve positioner, signal isolator);
   (iii) The candidate devices covered in the qualification report (this may include devices that failed the qualification), including the precise identification of the candidate devices by product name and version number of the software and hardware components;
   (iv)   Description of the technology used in the devices (which also implies the reason that the device is considered to be a smart device).

(b)   Reference to the plant requirements to be covered by the qualification. This may, for example, restrict the applicability to a specific plant or functions within a plant:

(i)   The specific target applications (safety functions) and the corresponding system class;

(ii)   The functional and performance requirements for the applications;

(iii)   The specification requirements for hardware environmental qualification (e.g. duration of the device mission under accident conditions);

(iv)   Maintainability and testability constraints (e.g. whether it is necessary for maintenance or testing to be possible without shutting down the plant);

(v)   The constraints on device selection based on plant architecture, including the possibilities of CCF with other layers of protection;

(vi)   The required safe failure modes and failure rates;

(vii)   The required device lifetime and duration of device support from the manufacturer;

(viii)   The required minimum level of QA programme of the designer and manufacturer.

(c)   Description of the candidate devices:

(i)   Manufacturers' description of the product or product line;

(ii)   Specific models, including software and hardware versions, addressed by the qualification, configuration and any other component or option that may pertain to the evaluation;

(iii)   References to product specification sheets and safety manuals;

(iv)   Specifics of programmable devices (i.e. microprocessors or HPDs) and the toolsets used to create the installed logic.

(d)   Summary of the qualification results:

(i)   Summary statement: successful or not;

(ii)   Safety class achieved;

(iii)   Applications covered by the qualification (or state if generic);

(iv)   Failure modes and failure rates;

(v)   Assumptions applied in the qualification (e.g. testing interval or operating temperature);

(vi)   Limitations on the qualification and restrictions on its use (as compared to the plant requirements);

(vii)   Documentation of any modifications needed to the product to achieve qualification;

(viii)   Modifications needed in the plant to accommodate the device;

(ix)   References to data sheets, calculations and site reports (or use attachments to the report).

(e)   Summary of gaps where the devices did not meet requirements, mitigation or compensatory activities executed or recommended and whether these activities close the gaps.

(f)   Special supplementary activities (such as tests and analyses by the licensee): list any additional activities required to achieve the necessary confidence in the device. If these have been executed, document the results of the activities and the degree of success in closing gaps. If these activities are yet to be executed, describe the minimum scope of the activities and the success criteria.

(g)   Data table: it can be helpful to users of the qualification report to produce a table summarizing the main assessment data, for example the following:
  (i)   Version numbers and configurations assessed;
  (ii)   Specific models of processors and HPDs and configuration tools used;
  (iii)   Performance characteristics;
  (iv)   Safe and dangerous failure rates and modes (e.g. fail high, fail as is);
  (v)   Systematic failure probability or capacity;
  (vi)   Proof test interval and possibility of on-line testing;
  (vii)   Maintenance requirements;
  (viii) Device lifetime and support duration.

(h)   A detailed report for each candidate device that contains a concise description of how the conclusion was reached and the evidence cited for each criterion, for example the following:
  (i)   Concise description of the device.
  (ii)   Compliance with conventional suitability requirements (clearly identify any requirements less than fully met):
           — Functional and performance requirements;
           — Interfacing with existing equipment (including a description of modifications to the device or other plant equipment that would be necessary);
           — Device lifetime and product support duration;
           — Average time to repair;
           — Testability while the plant is at power.
  (iii)   Hardware qualification test results (identify the laboratory and the specific test results):
           — Environmental;
           — Surge protection;
           — EMI/RFI;
           — Seismic.
  (iv)   Hardware reliability results (identify the analysis report and redundancy needed — if any).

(v) Software qualification results:
  — Primary function;
  — Non-interference of secondary or auxiliary functions with the primary function;
  — Resistance to cyberthreats;
  — User documentation for safety;
  — Design QA, V&V;
  — Configuration management.
(vi) Summary of deficiencies and compensatory methods applied.
(vii) Modifications required:
  — Modifications required to existing equipment or to plant staff training;
  — Modifications required to the smart device.
(viii) Recommendations to users.
(ix) References not easily available.

4.5. OTHER APPROACHES USED FOR SMART DEVICE QUALIFICATION

The detailed process for smart device qualification varies significantly among Member States, depending on regulatory requirements. Some Member States use third parties, in addition to the manufacturer qualification test, to conduct qualification tests for a sample of smart devices targeted for application in systems important to safety. Annex IV provides several examples of Member States' practices when qualifying smart devices for use in plant systems important to safety.

IAEA Nuclear Energy Series No. NR-T-3.31, Challenges and Approaches for Selecting, Assessing and Qualifying Commercial Industrial Digital Instrumentation and Control Equipment for Use in Nuclear Power Plant Applications [51], provides additional information regarding the justification and qualification processes for digital commercial products of limited functionality (equivalent to smart devices in this document) and further details regarding the following:

(a) The overall strategy for smart device qualification and its integration into the overall I&C safety justification;
(b) A step by step process for device qualification, including details on the expectations for various safety classes;
(c) Types of evidence to support the qualification;
(d) Competence required of the assessors.

Another source for qualification of a smart device for safe use in plant systems important to safety is IEC 62671 [3]. This standard has been specifically developed for selecting and using industrial DDLFs (i.e. smart devices) "that have not been produced to other IEC Standards which apply to systems and equipment important to safety in Nuclear Power Plants, but which are candidates for use in nuclear power plants" and it "provides requirements for the selection and evaluation of such devices where they have dedicated, limited, and specific functionality and limited configurability." At present, IEC 62671 [3] concentrates on individual cases of a smart device (or a family of such devices) in an NPP and includes only limited consideration of the possibility of more than one smart device and the consequent possibility of CCFs.

5. DEPLOYMENT OF A SMART DEVICE IN SYSTEMS IMPORTANT TO SAFETY

5.1. GENERAL

This section identifies several key aspects to consider over the life cycle of a smart device in an NPP, from the plant architecture and system design inputs to qualification, initial procurement, installation and commissioning, followed by operation so as to ensure the safe use of smart devices in systems important to safety and maintenance through the plant life.

5.2. CONFIGURATION MANAGEMENT

Requirement 10 of SSR-2/2 (Rev. 1) [38] states that "**The operating organization shall establish and implement a system for plant configuration management to ensure consistency between design requirements, physical configuration and plant documentation.**"

The plant's configuration management system controls the documents covering the characteristics of a facility's structures, systems and components (including computer systems and software) and ensures that changes to these characteristics are properly developed, assessed, approved, issued, implemented, verified, recorded and incorporated into the facility documentation.

The use of configuration management is a key component in all phases of introducing smart devices into an I&C architecture and explains how the software involved in smart devices (making smart device equipment complex) is subject to a number of influences. An uncontrolled configuration may lead to unacceptable behaviour under some plant conditions. Both detecting and preventing uncontrolled changes and the execution of controlled changes are strongly dependent on configuration management. Requirements on the application of a configuration management system to smart devices are established, for example, in IEC 61513 [52] and IEC 60880 [13].

Poor configuration management of smart devices may lead, for example, to the use of a different software version from the qualified version (e.g. in the case of software updates), potentially introducing operational risks.

For smart devices in NPPs, configuration management is used to record and control data such as the following:

— Plant configuration, including I&C architecture;
— Design basis of the plant I&C architecture;
— Design basis of each smart device in the plant, including the functional and performance specifications for the smart device;
— Locations within the architecture where a smart device is used;
— Details of each smart device in the plant, including all details of hardware, software and tools related to specific version identification and design, as well as V&V of the base (i.e. procured and qualified) version;
— User (plant operator) and maintainer documentation;
— Means to verify the installed version level of any software.

Configuration management programmes are normally established, maintained and followed by several entities, such as the manufacturer of each particular smart device, the I&C designer (NPP construction configuration management) and the licensee (NPP configuration management), and these entities have different motivations. It is essential to capture all necessary data within each specific configuration management system, especially since some of the entities provide source data for other entities.

5.3. SMART DEVICE LIFE CYCLE ACTIVITIES

5.3.1. Equipment selection

The smart devices to be procured and employed in a plant are initially selected according to the plant architecture design and individual system design. These activities result in inputs to the qualification process, such as the following:

— Identified common uses of the smart device in different layers of protection or in different channels;
— Level of safety significance;
— Functional and performance factors that could impact the selection of the specific device (e.g. range, adjustability, response time, failure rate);
— Required product lifetime.

A preliminary qualification (screening) may identify one or more devices that meet the functional requirements, but the need for detailed configuration management begins with the documentation specifying the device requirements.

5.3.2. Suitability assessment

The suitability assessment is a key step when using qualified devices. It consists of a verification that the qualified smart device is suitable for plant applications, considering their specific performance and safety requirements. The main aim of this activity is to determine whether the application fits the qualification envelope, that is, to consider whether there is any restriction of use or assumptions considered in the qualification that might make it unsuitable for the intended application. In cases where the qualification involves third party assessors, this step can also ensure that the licensee or duty holder takes ownership of the results of the qualification. Independent assessment or testing of the device may form part of this activity to provide additional assurance that the industrial or commercial grade device, which is not developed to nuclear standards, is suitable for the intended application.

5.3.3. Procurement

In the initial engagement with a smart device manufacturer, there are a number of aspects that need to be considered, including the following:

(a) Access to evidence of third parties: the willingness of the manufacturer to provide access to design documentation is a key success factor for the qualification process outlined in Section 4. Without positive cooperation between the assessors and the manufacturer, the likelihood of the justification being successful is low. The signing of a non-disclosure agreement early in the process is crucial, and the inclusion of third parties (such as the regulator in the framework) may need to be considered.

(b) Overall contractual arrangements: apart from normal contractual arrangements, there are several points that need to be defined for the supply of smart devices:

    (i) Feedback of failures from industrial customers;

    (ii) Supply chain management;

    (iii) Security implications;

    (iv) Obsolescence planning.

(c) Procurement strategy to manage obsolescence: this point is particularly crucial for smart devices for which version control (of both the product and the tools used to produce the product) is important. In some cases (e.g. a plant close to end of life), it could be sufficient to procure a large enough number of spares to cope with failures. In other cases (e.g. a new plant planned for 40–60 years of operation), it might be appropriate to agree on a time frame over which the qualified version will be available or on a framework for the qualification of any future version. In both cases, it is important that software support tools and their needed software environments be available for the foreseen period of use of the smart device type in the plant.

5.3.4. Installation and commissioning

Qualification provides a sufficiently high degree of assurance that a qualified device is suitable for installation. This needs to be verified for each specific device procured for installation or for spares and stores. Some of the elements considered in this section are also relevant for non-smart devices. The quality checks and version control become more challenging when software is involved, as visual checks are insufficient. Some considerations specific to installation and commissioning of smart devices are the following:

— Quality check of the incoming devices before installation (e.g. including disassembly of a device to identify any extraneous components);

— Version control before installation (e.g. using hashing techniques to verify the software version);

— Configuring security controls (often set to 'off' or 'none' by default in many smart devices).

Installation has to follow the manufacturer's installation manual, and the configuration for each instance of the smart device follows the set-up manual together with the specification arising from the plant design documentation. Commissioning tests are performed to exercise all safety related functions of the smart device. This may include simulating normal and accident conditions during the tests. Consideration needs to be given to testing to demonstrate that secondary functions do not interfere with the safety functions of the device. All these activities are performed using copies of the documents concerned retrieved from the configuration management system.

Another concern is that a device (smart or analogue) may contain an undeclared lower level digital component (e.g. FPGA element). Addressing this issue requires suitable verification during installation and commissioning and an effective supply chain management programme carried out by the manufacturer.

5.3.5. Plant operation using the smart device

The user documentation provided by the manufacturer needs to include a set-up manual and a user manual for plant maintainer and operator use. The latter will inevitably require customization to reflect the specific plant and the specific use of each instance of the smart device and parameterization (e.g. limits of operating range, type of response to operator input (possibly linear or square root)), and to describe the specific actions or operating procedures that the plant operators need to follow.

5.3.6. Periodic testing and maintenance

Periodic testing and maintenance are key to ensuring that smart devices are operating within the qualification boundaries. For example, periodic testing of a device, along with visual inspections of its connections and internal parts (where the device is not sealed), could identify signs of early ageing of a device caused by the environmental conditions in operation and failures not detectable by automatic internal diagnostics.

Key aspects of configuring and maintaining a smart device include ensuring the following:

— Training of personnel involved in commissioning, maintenance and testing.
— Quality of the commissioning and maintenance instructions for device qualification and parameterization.
— Availability of maintenance tools for the device configuration identified and justified in the qualification report.

— Monitoring and trending of failures or degradation of the devices (across the plant or across a fleet of plants). Note: the manufacturer may also maintain a database of field returns, but the NPP will have little control over the consistency of such data.

The points identified above are particularly challenging for those working with smart devices compared with analogue components because of the additional complexity of smart devices (e.g. because of their configurability) and the skill set required for these activities (e.g. a combined hardware and software background), as well as the need to manage cybersecurity risks. This can be achieved by establishing processes and procedures to verify whether there has been tampering with the device or its configuration (including through tools used for its configuration). An adequate process for managing cybersecurity configurations includes password management and software updates (if applicable).

5.4. MANAGEMENT OF CHANGE

This section addresses the source and implementation of changes affecting the NPP and smart devices. The need to make modification to the smart device may result, for example, from one of the following sources:

— A device replacement because of failure or ageing (requiring configuration);
— A replacement device not containing the correct software version;
— A software update from the smart device vendor (e.g. fixing software flaws or patching) that requires upload of a new version;
— Changes made to a plant architecture or system that comprises one or more embedded smart devices.

Typically, these types of modifications affect the validity of the smart device qualification, either because the device to be implemented may be a different version (resulting from software or hardware changes) or because of modifications in the plant operational requirements.

The impact of smart device changes on the original device qualification depends on the changes in the plant requirements or modifications to software and hardware. In order to determine the scope of the review of the original qualification, the impact assessment needs to consider the following:

— The precise identification of the pre-modification and post-modification smart device;

— Whether the modification plan reflects the manufacturer's development life cycle;
— Whether the personnel involved in the modification are at least as competent as those involved in the original design, including use of the same tools;
— Whether the verification plan covers all functionality of the smart device;
— The complete documentation of the software before and after modification;
— The approval of the change by a committee including subject matter experts;
— The implementation of the modified smart device to the plant in a way analogous to the original installation and commissioning;
— The relevance of the evidence assessed as part of the previous qualification;
— Hardware changes, such as materials used, precision in dimensioning and finishing;
— Changes in the development and manufacturing processes of the vendor;
— Changes in the supply chain;
— Rigour of testing.

The assessment of the impact of changes on the smart device needs to examine the following as a minimum:

— The impact of not making the change;
— The possible impact of faults introduced while making the change;
— The skill level required to make the change;
— The competence of the modifying organization to make the change;
— How the change can be verified and how the qualification can be updated.

Operational experience or safety alerts could provide indications of non-conformities that may require revisiting the original smart device qualification and making changes in the scope or using additional evidence supporting the qualification, as necessary.

REFERENCES

[1]   INTERNATIONAL ELECTROTECHNICAL COMMISSION, Functional Safety of Electrical/Electronic/Programmable Electronic Safety-Related Systems, IEC 61508, IEC, Geneva (2010).

[2]   INTERNATIONAL ATOMIC ENERGY AGENCY, Design of Instrumentation and Control Systems for Nuclear Power Plants, IAEA Safety Standards Series No. SSG-39, IAEA, Vienna (2016).

[3]   INTERNATIONAL ELECTROTECHNICAL COMMISSION, Nuclear Power Plants, Instrumentation and Controls Important to Safety, Selection and Use of Industrial Digital Devices of Limited Functionality, IEC 62671, IEC, Geneva (2013); Corrigendum IEC 62671/COR1 (2016).

[4]   STANDARDIZATION ASSOCIATION FOR MEASUREMENT AND CONTROL IN CHEMICAL INDUSTRIES (NAMUR), Standardization of the Signal Level for the Failure Information of Digital Transmitters, Standard NE43, NAMUR, Leverkusen, Germany (2021).

[5]   ASTM INTERNATIONAL, Standard Guide for the Measurement of Single Event Phenomena (SEP) Induced by Heavy Ion Irradiation of Semiconductor Devices, ASTM F1192-11, West Conshohocken, PA (2018).

[6]   EUROPEAN SPACE AGENCY, Single Event Effects Test Method and Guidelines, ESCC Basic Specification No. 25100, ESA, Paris (2014).

[7]   JOINT ELECTRON DEVICE ENGINEERING COUNCIL, Test Procedure for the Management of Single-Event Effects in Semiconductor Devices from Heavy Ion Irradiation, JESD57A, JEDEC, Arlington, VA (2017).

[8]   UNITED STATES DEPARTMENT OF DEFENSE, Test Methods for Semiconductor Devices, MIL-STD-750F_CHG-2, US DOD, Washington, DC (2016).

[9]   LAWRENCE LIVERMORE NATIONAL LABORATORY, Method for Performing Diversity and Defence-in-Depth Analyses of Reactor Protection Systems, US NRC NUREG/CR-6303, LLNL, Livermore, CA (1994).

[10]   NUCLEAR REGULATORY COMMISSION, Guidance for Evaluation of Diversity and Defense-in-Depth in Digital Computer-Based Instrumentation and Control Systems, Branch Technical Position 7-19, NRC, Washington, DC (2016).

[11]   VDI/VDE-GESELLSCHAFT MESS- UND AUTOMATISIERUNGSTECHNIK, Design Criteria Serving to Ensure Independence of I&C Safety Functions, VDI/VDE 3527, VDI, Düsseldorf (2002).

[12]   INTERNATIONAL ELECTROTECHNICAL COMMISSION, Nuclear Power Plants, Instrumentation and Control Systems Important to Safety, Requirements for Coping with Common Cause Failure (CCF), IEC 62340, IEC, Geneva (2007).

[13]   INTERNATIONAL ELECTROTECHNICAL COMMISSION, Nuclear Power Plants, Instrumentation and Control Systems Important to Safety, Software Aspects for Computer-Based Systems Performing Category A Functions, IEC 60880, IEC, Geneva (2006).

[14] INSTITUTE OF ELECTRICAL AND ELECTRONIC ENGINEERS, IEEE Standard Criteria for Safety Systems for Nuclear Power Generating Stations, IEEE 603, IEEE, New York (2018).

[15] INSTITUTE OF ELECTRICAL AND ELECTRONIC ENGINEERS, IEEE Standard Criteria for Programmable Digital Devices in Safety Systems of Nuclear Power Generating Stations, IEEE 7-4.3.2, IEEE, New York (2016).

[16] CANADIAN STANDARDS ASSOCIATION, Qualification of Digital Hardware and Software for Use in Instrumentation and Control Applications for Nuclear Power Plants, CSA N290.14-15 (R2020), CSA Group, Toronto (2020).

[17] ELECTRIC POWER RESEARCH INSTITUTE, Guideline for the Acceptance of Commercial-Grade Items in Nuclear Safety-Related Applications: Revision 1 to EPRI NP-5652 and TR-102260, TR-3002002982, EPRI, Palo Alto, CA (2014).

[18] ELECTRIC POWER RESEARCH INSTITUTE, Guideline on Evaluation and Acceptance of Commercial Grade Digital Equipment for Nuclear Safety Applications, TR-106439, EPRI, Palo Alto, CA (1996).

[19] ELECTRIC POWER RESEARCH INSTITUTE, Generic Requirements Specification for Qualifying a Commercially Available PLC for Safety-Related Applications in Nuclear Power Plants, TR-107330, EPRI, Palo Alto, CA (1996).

[20] NUCLEAR REGULATORY COMMISSION, Dedication of Commercial-Grade Items for Use in Nuclear Power Plants, Regulatory Guide 1.164, NRC, Washington, DC (2017).

[21] ELECTRIC POWER RESEARCH INSTITUTE, Guideline for the Utilization of Commercial Grade Items in Nuclear Safety Related Applications (NCIG-07), NP-5652, EPRI, Palo Alto, CA (1998).

[22] ELECTRIC POWER RESEARCH INSTITUTE, Guidance for Commercial Grade Dedication, TR-102260, EPRI, Palo Alto, CA (2011).

[23] FRENCH ASSOCIATION FOR DESIGN, CONSTRUCTION AND IN-SERVICE INSPECTION RULES FOR NUCLEAR ISLAND COMPONENTS, Design and Construction Rules for Electrical and I&C Systems and Equipment, RCC-E, AFCEN, Paris (2016).

[24] OFFICE FOR NUCLEAR REGULATION, Safety Assessment Principles for Nuclear Facilities, 2014 Edition, Revision 1 (January 2020), ONR CM9 Ref 2019/367414, ONR, Bootle (2020).

[25] CANADIAN STANDARDS ASSOCIATION, Quality Assurance Program Requirements for the Supply of Items and Services for Nuclear Power Plants, CSA Z299.3, CSA, Toronto (2006).

[26] NUCLEAR REGULATORY COMMISSION, Title 10 Code of Federal Regulations (CFR) Part 50, Appendix B, Quality Assurance Criteria for Nuclear Power Plants and Fuel Reprocessing Plants, NRC, Washington, DC (2019),
https://www.nrc.gov/reading-rm/doc-collections/cfr/part050/part050-appb.html

[27] AMERICAN SOCIETY OF MECHANICAL ENGINEERS, Quality Assurance Requirements for Nuclear Facility Applications, ASME NQA-1-2015, ASME, New York (2015).

[28] NUCLEAR REGULATORY COMMISSION, Title 10 Code of Federal Regulations (CFR) Part 50, Part 50.55a, Codes and Standards, NRC, Washington, DC (2019),
https://www.nrc.gov/reading-rm/doc-collections/cfr/part050/part050-0055a.html

[29] NUCLEAR REGULATORY COMMISSION, Quality Assurance Program Criteria (Design and Construction), Regulatory Guide 1.28, NRC, Washington, DC (2017).

[30] AMERICAN SOCIETY OF MECHANICAL ENGINEERS, Quality Assurance Requirements for Nuclear Facility Applications Addenda 1a, ASME NQA-1a-2008, ASME, New York (2008).

[31] AMERICAN SOCIETY OF MECHANICAL ENGINEERS, Quality Assurance Requirements for Nuclear Facility Applications, ASME NQA-1-2012, ASME, New York (2012).

[32] INSTITUTE OF ELECTRICAL AND ELECTRONIC ENGINEERS, IEEE Standard for System, Software, and Hardware Verification and Validation, IEEE 1012, IEEE, New York (2016).

[33] INSTITUTE OF ELECTRICAL AND ELECTRONIC ENGINEERS, IEEE Standard for Configuration Management in Systems and Software Engineering, IEEE 828, IEEE, New York (2012).

[34] KAERI, KINS, NSC, ONR, SSM, STUK, Licensing of Safety Critical Software for Nuclear Reactors, Common Position of International Regulators and Authorised Technical Support Organizations, Revision 2018, Task Force on Safety Critical Software (2018),
http://www.onr.org.uk/software.pdf

[35] INTERNATIONAL ATOMIC ENERGY AGENCY, Protecting Against Common Cause Failures in Digital I&C Systems of Nuclear Power Plants, Nuclear Energy Series No. NP-T-1.5, IAEA, Vienna (2009).

[36] INTERNATIONAL ATOMIC ENERGY AGENCY, Computer Security at Nuclear Facilities, IAEA Nuclear Security Series No. 17 (Rev. 1), IAEA, Vienna (2021).

[37] INTERNATIONAL ATOMIC ENERGY AGENCY, Computer Security of Instrumentation and Control Systems at Nuclear Facilities, IAEA Nuclear Security Series No. 33-T, IAEA, Vienna (2018).

[38] INTERNATIONAL ATOMIC ENERGY AGENCY, Safety of Nuclear Power Plants: Design, IAEA Safety Standards Series No. SSR-2/1 (Rev. 1), IAEA, Vienna (2016).

[39] INTERNATIONAL ELECTROTECHNICAL COMMISSION, Nuclear Power Plants, Instrumentation and Control Systems Important to Safety, Software Aspects for Computer-Based Systems Performing Category B or C Functions, IEC 62138, IEC, Geneva (2018).

[40] INTERNATIONAL ATOMIC ENERGY AGENCY, Equipment Qualification for Nuclear Installations, IAEA Safety Standards Series No. SSG-69, IAEA, Vienna (2021).

[41] INTERNATIONAL ELECTROTECHNICAL COMMISSION, Nuclear Facilities, Electrical Equipment Important to Safety, Qualification, IEC/IEEE 60780-323 (Edition 1.0), IEC, Geneva (2016).

[42] INTERNATIONAL ATOMIC ENERGY AGENCY, Seismic Design for Nuclear Installations, IAEA Safety Standards Series No. SSG-67, IAEA, Vienna (2021).

[43]   INTERNATIONAL ELECTROTECHNICAL COMMISSION, IEEE/IEC International Draft Standard, Nuclear Facilities, Equipment Important to Safety, Seismic Qualification, IEC/IEEE 60980-344, IEC, Geneva (2020).

[44]   INTERNATIONAL ELECTROTECHNICAL COMMISSION, Electromagnetic Compatibility (EMC) Product Family Standards: Emission, IEC 61000-3/4/6, IEC, Geneva (2018).

[45]   UNITED STATES DEPARTMENT OF DEFENSE, Requirements for the Control of Electromagnetic Interference Characteristics of Subsystems and Equipment, MIL-STD-461G, US DOD, Washington, DC (2015).

[46]   INSTITUTE OF ELECTRICAL AND ELECTRONIC ENGINEERS, IEEE Recommended Practice on Characterization of Surges in Low-Voltage (1000 V and Less) AC Power Circuits, IEEE C62.41.2, IEEE, New York (2003); Corrigendum IEEE C62.41.2/COR 1 (2012).

[47]   INTERNATIONAL ELECTROTECHNICAL COMMISSION, Industrial Communication Networks, Network and System Security, Part 3-3: System Security Requirements and Security Levels, IEC 62443-3-3, IEC, Geneva (2013).

[48]   INTERNATIONAL ATOMIC ENERGY AGENCY, Leadership and Management for Safety, IAEA Safety Standards Series No. GSR Part 2, IAEA, Vienna (2016).

[49]   INTERNATIONAL ATOMIC ENERGY AGENCY, Application of the Management System for Facilities and Activities, IAEA Safety Standards Series No. GS-G-3.1, IAEA, Vienna (2006).

[50]   INTERNATIONAL ATOMIC ENERGY AGENCY, The Management System for Nuclear Installations, IAEA Safety Standards Series No. GS-G-3.5, IAEA, Vienna (2009).

[51]   INTERNATIONAL ATOMIC ENERGY AGENCY, Challenges and Approaches for Selecting, Assessing and Qualifying Commercial Industrial Digital Instrumentation and Control Equipment for Use in Nuclear Power Plant Applications, IAEA Nuclear Energy Series No. NR-T-3.31, IAEA, Vienna (2020).

[52]   INTERNATIONAL ELECTROTECHNICAL COMMISSION, Nuclear Power Plants, Instrumentation and Control Systems Important to Safety, General Requirements for Systems, IEC 61513, IEC, Geneva (2016).

Annex I

ADDITIONAL CONSIDERATIONS ON THE USE OF SMART DEVICES

This annex provides suggestions for additional criteria for device selection and use of the device once it has been qualified.

I–1. SUGGESTED ADDITIONAL SELECTION CRITERIA

Suggested additional selection criteria include limiting the selection to devices that use technologies with the following characteristics:

— Are proven by several years of widespread use, e.g. a chip whose manufacturing technology (e.g. 10 nm) and toolset have been in use for three years;
— Have a high likelihood of being supported by the manufacturer for a good fraction of the remaining life of the plant.

In addition, the selection can be limited to vendors that meet the following criteria:

— Have already obtained accredited certification of their product to a recognized international safety standard, such as IEC 61508 [I–1];
— Are willing to support the specific selected product version for a guaranteed time and will warn the plant before such support is suspended, so that a supply of spares can be procured;
— Maintain a web site or other means to allow users to report and inquire about problems with the product or its components;
— Are willing to put the design into escrow as a guarantee that the plant will have access to the detailed design information if the manufacturer is not able to support it;
— Are willing to make specific design changes to suit the needs of the plant and to make them so as to preserve pre-existing certification of the product.

I–2. SUGGESTED ADDITIONAL CONSIDERATIONS DURING INTEGRATION AND USE

Additional approaches that may be considered during integration and use of smart devices include the following:

— Involve operations in human factors for use of the smart device (and possibly selection as well);
— Maintain a database of all smart devices in the plant, complete with low level details of processor models, manufacturers and specific versions of tools used to program and configure the device (consulted to check for possible common cause failures when qualifying any new device to introduce into the plant);
— Maintain a database of failures of all smart devices in the plant;
— Plan a maintenance strategy based on the capabilities at the plant;
— Consider a 'repair by replacement' strategy whereby faulty devices are sent to the manufacturer for repair and replaced with a spare;
— Collect all necessary design, operation and maintenance information at the end of a device lifetime to support finding an adequate replacement.

REFERENCE TO ANNEX I

[I–1] INTERNATIONAL ELECTROTECHNICAL COMMISSION, Functional Safety of Electrical/Electronic/Programmable Electronic Safety-Related Systems, IEC 61508, IEC, Geneva (2010).

Annex II

EXAMPLE OF COMMON CAUSE FAILURE ANALYSIS

This annex presents some examples of how the use of smart devices at different layers in the overall instrumentation and control (I&C) architecture can introduce common cause failure (CCF) vulnerabilities that need consideration in safety applications.

II–1. OVERVIEW OF SMART DEVICE ARCHITECTURE AND INTERNAL ELEMENTS

As defined in Section 1, a smart device is a digital component with limited functionality that is configurable but not reprogrammable by the end user. Even the simplest smart device can be quite complex in its internal architecture and in the layered software structure. Figure II–1 provides a schematic representation of the key hardware components in a smart device, and Fig. II–2 provides an example of software components used in the design, operation and configuration of a smart device.[1] Figures II–1 and II–2 highlight the interconnection and complexity of the architecture of even a simple smart device and help illustrate the challenge in adequately controlling the risks deriving from the use of a commercial device for a high integrity or safety application.

Figures II–1 and II–2 help identify various aspects of interest for CCF analysis, such as the following:

— There are a number of internal components in a smart device, each of which may (or may not) itself be digitally reconfigurable. The internal architecture of a smart device depends on the design and technology solution chosen by the manufacturer, although it typically contains a number of software elements (e.g. application software, operating system, libraries, predefined function blocks).

---

[1] Examples of hardware elements in a typical smart device are microprocessors, field-programmable gate arrays, discrete electronics, input/output ports, the watchdog and power supplies. Examples of software elements in a typical smart device include the operating system, application software and other pre-developed software (e.g. function blocks, libraries, protocols). Examples of software tools used in smart device development are configuration tools/software, the compiler/linker and other testing and analysis tools.

— The systematic failures potentially affecting smart devices are related not only to the firmware and software installed in the smart device itself, but also to the tools used in its design and configuration.

— Different types and models of smart devices and different manufacturers may use the same components or software modules, making them susceptible to CCFs even if they are not identical. This has to be considered when addressing CCF concerns at plant architectural level.

II–2. TRIGGERING COMMON CAUSE FAILURE AMONG SMART DEVICES

Software flaws in smart devices can cause their failure. When the same device is implemented at different levels in the plant architecture, this can result in a CCF. Figure II–3 illustrates how this can occur through the failure propagation model from Ref. [II–1] and identifies different elements that can contribute to a potential CCF:

(a) Context: this element is outside the individual smart devices and typically depends on the plant condition and the initiating event. Architectural decisions regarding which physical parameters are fed into smart devices in the overall plant architecture can contribute to the triggering of smart device failures (possibly multiple failures).

(b) Device level: this includes all the elements that determine the behaviour of a smart device, such as:

(i) Device state: this is specific to each smart device and comprises the internal state of the device, as well as the external operating context (e.g. environmental conditions). Among other factors, this can be influenced by the maintenance regimes.

(ii) Software flaws: these are typically residual errors, weaknesses or vulnerabilities not identified or adequately mitigated against in development or testing. It is worth noting that, because of the various levels of software involved in a smart device (see Fig. II–2), the same types of flaws can affect devices of different models (e.g. development tools or operating systems) and even different manufacturers.

Figure II–3 provides the following two examples of how CCFs may affect multiple smart devices:

(a) Failure of device 1 (model A) and device 2 (model B): in this example, devices 1 and 2 receive the same input (e.g. a pressure signal from the

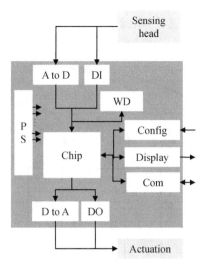

FIG. II–1. *Typical elements of hardware and software in a smart device. A to D: analogue to digital; D to A: digital to analogue; DI: digital input; PS: power supply; DO: digital output; WD: watchdog; Config: configuration interface; Com: communication module.*

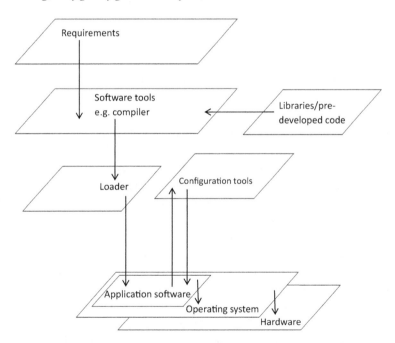

FIG. II–2. *Example of a layered structure of digital elements used in the design and configuration of a smart device.*

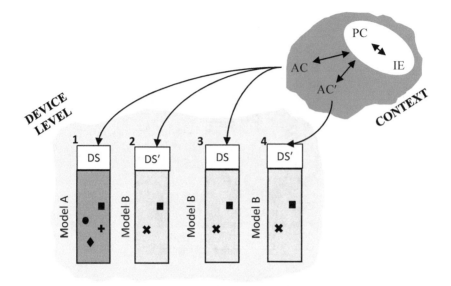

FIG. II–3. Propagation model for failure of multiple smart devices used in a nuclear power plant. IE: initiating event (e.g. loss of coolant); PC: plant conditions at the moment of occurrence of the IE (e.g. general operating conditions and maintenance); AC, AC': signal trajectories caused by the IE (e.g. temperature evolution in one of the steam generators); DS, DS': device states, including software internal states (e.g. memory states) and operating conditions (e.g. environmental conditions, maintenance and calibration) at the moment of occurrence of the IE. The symbols represent residual errors, weaknesses or vulnerabilities not identified or adequately mitigated during development and testing.

primary circuit). In the example, it is assumed that the devices are in different locations in the plant and subject to different maintenance and operating conditions (DS and DS' in the figure). Although different models, it is possible that they could exhibit an unexpected behaviour in the same time window, for example in the case of an anomaly in the signal trajectory (not expected in the requirement specification) or a common element of the software being present in both devices, such as the same software libraries or real time operating system being used in different models.

(b) Failure of device 3 (model B) and device 4 (model B): in this example, devices 3 and 4 receive different input signals (AC and AC' in the figure, which could be, for example, pressure in the primary circuit and pressure in the secondary circuit). Because the device model is the same and the two devices are subject to similar operating conditions (DS in the figure), there is a potential for common vulnerabilities to cause the device to stop working in the same time window (e.g. susceptibility to power supply fluctuations).

II–3. COMMON CAUSE FAILURE ANALYSIS

Figure 1 in Section 3 shows a CCF analysis step in the consideration of the device selection and confirmation of the robustness of the architecture to CCF. The approach to CCF analysis may vary depending on the practices applied by different Member States, but it is typically expected to cover the following:

(a) Consideration of the risk associated with multiple smart device failures: the adequacy of the architectural solution has to be considered in the context of the risk associated with CCFs.[2] In this context, two elements have to be considered:

  (i) Consequences: this is the impact of the CCF in terms of nuclear safety (e.g. radiological releases). The deterministic safety analyses used to derive requirements for the architecture may need to be confirmed or complemented considering the specific failure modes of smart devices. For example, particular attention needs to be given to spurious actuation, where failure of a smart device model can both initiate an accident scenario and simultaneously defeat multiple defences.

  (ii) Likelihood: this is related to the probability of occurrence of CCFs and the correlation of multiple failures. There is generally significant uncertainty on the estimation of the correlation factor among smart devices, as well as difficulty in estimating software reliability. For this reason, in many Member States, it is generally preferable to take a deterministic approach, assume that a CCF among devices can happen and focus attention on consequences. If a probabilistic argument is used as part of the justification, sensitivity analyses are always recommended to account for uncertainty in the correlation between failures and cliff edge effects.

(b) Analysis of the variety of conditions that could trigger multiple device failures: these are related to the existence of design flaws in the smart device, which can result in CCFs if triggered in a limited time window. With reference to Fig. II–4, the triggering mechanisms are mainly related to the following:

  (i) Activation conditions: a set of smart devices could be subject to the same input (e.g. primary circuit temperature measured in different loops) or the same physical quantity (e.g. pressure in the pressurizer).

---

[2] In some Member States (e.g. the UK), the expectation is that the risk associated with CCF is reduced to as low as reasonably practicable. See www.onr.org.uk/documents/tolerability.pdf

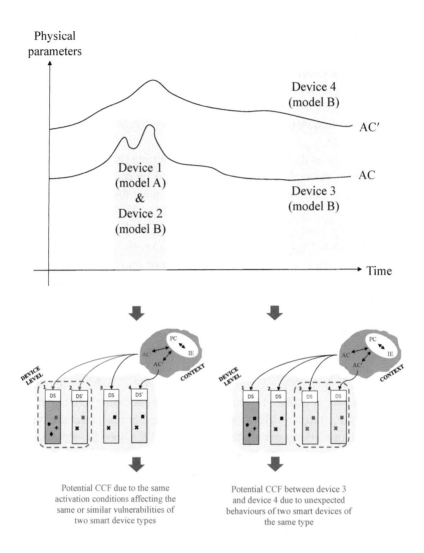

Physical parameters

Device 4
(model B)

AC′

Device 1
(model A)
&
Device 2
(model B)

AC

Device 3
(model B)

Time

Potential CCF due to the same activation conditions affecting the same or similar vulnerabilities of two smart device types

Potential CCF between device 3 and device 4 due to unexpected behaviours of two smart devices of the same type

FIG. II–4. Examples of common cause failure among smart devices.

Referring to Fig. II–4, this could generate the potential for a CCF across multiple devices and needs to be considered in the CCF analysis.

(ii) Device state: other failures that cannot be directly related to the application software of the smart device can be associated with the operating system or memory management. Operational and maintenance practices may affect the potential for CCF of multiple devices and need be considered in the analyses. For example, maintenance errors may cause multiple devices to fail (e.g. by failing to use a properly calibrated instrument, injecting the wrong value for testing or erroneously changing configuration parameters). In addition, environmental conditions may cause problems in digital systems (e.g. electromagnetic interference or radiofrequency interference), such as accelerated ageing due to temperatures being higher than expected or electrical input disturbance causing undesired behaviour of a device.

The CCF analysis needs to cover a comprehensive list of triggering mechanisms. For example, during retrofitting the existing documentation may not identify all potential triggers relevant to smart devices (e.g. existing analogue equipment may not be affected in the same manner as a replacement smart device when exposed to electromagnetic or radiofrequency interference in the plant operating environment). Additional analysis may be required to ensure that all potential triggers have been identified, their consequences understood and the risks adequately mitigated.

II–4. EXAMPLES OF USE OF SMART DEVICES IN THE INSTRUMENTATION AND CONTROL ARCHITECTURE

This section presents some examples of how the use of smart devices at different levels in the overall I&C architecture can introduce vulnerabilities that need consideration in safety applications.

II–4.1. Use of the same smart device at different levels of defence in depth

The key consideration here is to meet Requirement 7 of SSR-2/1 (Rev. 1) [II–2]: "The levels of defence in depth shall be independent as far as is practicable".

The examples shown in Figs II–5 and II–6 outline different scenarios and challenges in the justification of an I&C architectural solution.

Example 1 (Fig. II–5)

— I&C architecture: the same smart pressure transmitter type is used both for prevention of abnormal operation (Level 2 of defence in depth) and to protect against a certain fault sequence (Level 3).
— Scenario: a spurious signal from a smart device (e.g. low pressure) can induce a transient in the plant. The same transmitter could then be needed to support another safety function in the progress of the accident scenario.

Defence in depth level	Objective	Example
Level 1	To prevent deviations from normal operation and the failure of items important to safety	1
Level 2	To detect and control deviations from normal operational states in order to prevent anticipated operational occurrences at the plant from escalating to accident conditions	2 3
Level 3	To prevent damage to the reactor core or to prevent radioactive releases requiring off-site protective actions; to return the plant to a safe state	
Level 4	To mitigate the consequences of accidents that result from failure of the third level of defence in depth	
Level 5	To mitigate the radiological consequences of radioactive releases that could potentially result from accidents	

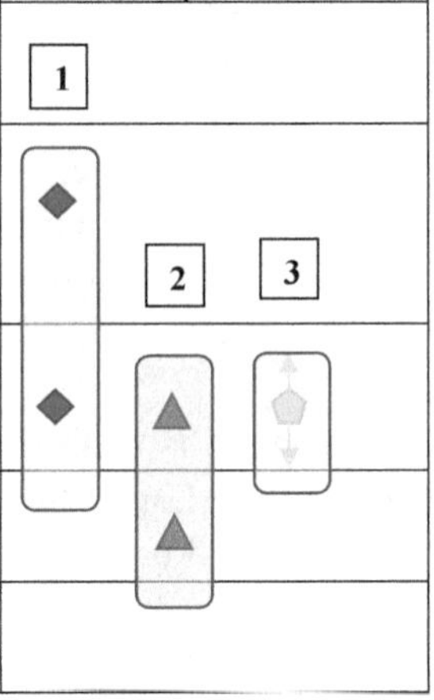

FIG. II–5. Examples of using the same smart device in different layers of defence. Objectives are defined according to SSR-2/1 (Rev. 1) [II–2].

— Implications: if both transmitters become simultaneously faulty as a result of CCF, the plant might not be within the boundary of the safety demonstration. A more detailed analysis might be required to understand the following:
  • How dangerous failure modes may affect both smart devices;
  • The suitability of other protection in place to cope with the accident scenario (e.g. relying on a different process variable or different devices).

Example 2 (Fig. II–5)

— I&C architecture: the same pressure transmitter is used both in design basis defence (Level 3 of defence in depth) and severe accident defence (Level 4).
— Scenario: in an accident scenario, failure of the pressure measurement system could cause the plant to reach severe accident conditions. In such a case, the unavailability of a second pressure transmitter could affect the robustness of the severe accident provisions.
— Implications: in this case, the loss of independence between layers may not be acceptable ('Swiss cheese' model; see Ref. [II–3]) and the use of diversity may be considerably more robust.

Example 3 (Fig. II–5)

— I&C architecture: the same device provides a signal to indicate normal operation of the I&C system (Level 2 in the defence in depth) and accident condition of the I&C system (Level 3).
— Scenario: in this case, a single failure of a device (without CCF) could cause multiple failures in the overall I&C architecture.
— Implications: the lack of signal segregation in the overall I&C architecture may need to be reviewed.

II–4.2. Use of the same smart device in different redundancies of the same system

The use of redundancies (also called channelization) is intended to increase the availability of the safety function. The key focus here is to consider how the use of various smart devices of the same type in redundancies of the same I&C system affects the system reliability.

Example 1 (Fig. II–6)

— I&C architecture: the same pressure transmitter is used in four redundancies of an I&C system (2oo4 voting logic). Two or more transmitters are needed to ensure the success of this logic.
— Scenario: CCF of two (or more) transmitters in the four redundancies could impair the ability of the system to deliver a safety function with the required reliability. This could also result from a combination of an unrevealed

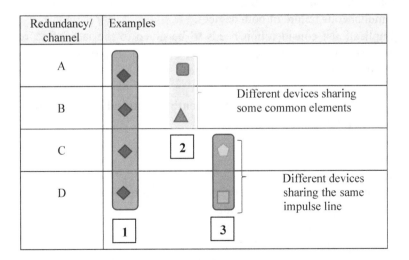

FIG. II–6. Examples of using the same smart device in different redundancies.

failure and a second failure on demand on a different device of the same type, rendering the system unable to take action when required.

— Implications: considering full correlation of the failures, the reliability of the system is limited by the reliability of the single device. Depending on the relevance of the safety function, consideration could be given to having two diverse sets of transmitters, which would improve the resilience of the system (no single CCF could affect the 2oo4 voting).

Example 2 (Fig. II–6)

— I&C architecture: two configurations of the same transmitter (e.g. pressure and differential pressure) or two transmitters using a common functionality (e.g. highway addressable remote transducers) are used in different redundancies of an I&C system.
— Scenario: although not identical in make and use, there is a potential for these smart devices to be affected by a coincident failure.
— Implications: the level of diversity between the two devices and the differences in use need to be considered to determine the adequacy of the I&C architecture.

Example 3 (Fig. II–6)

— I&C architecture: the tap-off point and related impulse piping are shared between two different instruments.
— Scenario: failure or clogging of the impulse piping could cause the simultaneous failure of both devices.
— Implications: consideration needs to be given to the lack of segregation between the lines and the impact of a CCF. This is a particular problem if the fluid involved contains particulates.

II–4.3. Use of smart devices in the essential electrical distribution system architecture

Example 1

— Electrical architecture: diesel generators are provided in multiple divisions to supply loads important to safety.
— Scenario: CCF in smart devices embedded in the diesel generators results in the loss of standby generation in all divisions.
— Implications: this could result in the failure to start of all the standby generation in the plant following loss of off-site power, leading to a station

blackout. This could be mitigated by the provision of a diesel generator in at least one division operating at a lower voltage level using different technologies. An alternative mitigation method would be for a diesel generator in one division to be designed without the use of smart devices, which could limit its functionality but would provide protection against CCF. These measures would ensure the capability of supplying essential loads in order to provide resilience and maintain the plant in a controlled state.

Example 2

— Electrical architecture: a plant monitoring system connects the plant protection and monitoring devices to a display in a central control room.
— Scenario: CCF is introduced into multiple protection relays from the monitoring system, resulting in loss of electrical supplies caused by tripping of circuit breakers.
— Implications: this could result in the total loss of all electrical supplies in the plant. This could be mitigated by using a monitoring system architecture that communicates with plant relays in read-only mode for monitoring, with no capability to operate plant items.

Example 3

— Electrical architecture: the electrical equipment (e.g. rectifier) has a smart device as a control unit.
— Scenario: CCF of smart devices could result in failure of electrical protection relays to operate in electrical fault conditions. This would be mitigated by the use of upstream smart devices, provided that these are not identical. Another failure mode could result in spurious tripping, which could lead to loss of power supplies.
— Implications: the implications of CCF for electrical equipment are dependent on the failure modes of the individual items. There is a need to perform a safety assessment to consider the consequences of each failure mode and the architectural measures that can be taken in response to this safety assessment.

Example 4

— Electrical architecture: the electrical distribution system consists of identical redundant divisions.

— Scenario: where identical smart devices are installed in redundant divisions, there is the potential for CCF to result in the loss of power supplies to multiple divisions of equipment.
— Implications: loss of multiple divisions could result in a station blackout condition. This could be mitigated by the provision of equipment diversity by utilizing equipment from different manufacturers.

Example 5

— Electrical architecture: many items of the electrical equipment contain smart devices.
— Scenario: two or more identical smart devices fail simultaneously on a common cause.
— Implications: as CCF of electrical equipment can result in the loss of supply to safety systems, the option of providing battery supplies to essential equipment required to maintain the plant in a controlled state needs to be considered.

REFERENCES TO ANNEX II

[II–1] NUCLEAR ENERGY AGENCY, COMMITTEE ON THE SAFETY OF NUCLEAR INSTALLATIONS, Failure Modes Taxonomy for Reliability Assessment of Digital Instrumentation and Control Systems for Probabilistic Risk Analysis, NEA/CSNI/R(2014)16, NEA, Paris (2015).
[II–2] INTERNATIONAL ATOMIC ENERGY AGENCY, Safety of Nuclear Power Plants: Design, IAEA Safety Standards Series No. SSR-2/1 (Rev. 1) IAEA, Vienna (2016).
[II–3] REASON, J., The contribution of latent human failures to the breakdown of complex systems, Phil. Trans. R. Soc. Lond. B **327** (1990) 475–484.

Annex III

USE OF STANDARDS AND GUIDANCE FOR SOFTWARE QUALIFICATION

IEC 62671 [III–1] is the International Electrotechnical Commission (IEC) standard for qualifying digital devices of limited functionality (DDLFs). IEC 61508 [III–2] is the most widely used safety standard under which DDLFs are certified for non-nuclear applications. The other standards in Tables III–1 and III–2 provide the requirements or guidance for design according to nuclear standards, and thus offer useful information relative to qualification objectives.

TABLE III–1. INTERNATIONAL STANDARDS THAT HAVE A STRONG RELATIONSHIP WITH THIS SAFETY REPORT

IEC 62671/COR1 [III–3]	Nuclear power plants — instrumentation and control important to safety — selection and use of industrial digital devices of limited functionality; Corrigendum 1
	This standard has been specifically developed for selecting and using DDLFs It is intended to be applied to DDLFs that have not been produced to other IEC standards that apply to systems and equipment important to safety in nuclear power plants (NPPs), but are candidates for use in NPPs. It addresses certain devices that contain embedded software or electronically configured digital circuits that have not been produced to other IEC standards that apply to systems and equipment important to safety in NPPs, but are candidates for use in NPPs. It provides requirements for the selection and evaluation of such devices where they have dedicated, limited and specific functionality, and limited configurability.
IEC 61508[a] [III–4]	Functional safety of electrical/electronic/programmable electronic safety-related systems
	This is an industrial (non-nuclear) standard that is accepted in some Member States for use with smart devices. It covers those aspects to be considered when electrical/electronic/programmable electronic systems are used to carry out safety functions. A major objective of this standard is to facilitate the development of international standards for the product and application sector by the responsible technical committees.

TABLE III–1. INTERNATIONAL STANDARDS THAT HAVE A STRONG
RELATIONSHIP WITH THIS SAFETY REPORT (cont.)

IEC 61513
[III–5]

Nuclear power plants — instrumentation and control important to
safety — general requirements for systems

This standard is used in some Member States for guidance on
integration of smart devices in the overall instrumentation and
control (I&C) systems of NPPs. It provides requirements and
recommendations for the overall I&C architecture, which may
contain either or both technologies.

IEC 60880
[III–6]

Nuclear power plants — instrumentation and control systems
important to safety — software aspects for computer-based systems
performing Category A functions

This standard provides requirements for the software of
computer based I&C systems of NPPs performing functions of
safety Category A, as defined by IEC 61226 [III–7]. It provides
requirements for the purpose of achieving highly reliable software
and addresses each stage of software generation and documentation,
including requirements specification, design, implementation,
verification, validation and operation.

IEC 60987
[III–8]

Nuclear power plants — instrumentation and control important to
safety — hardware design requirements for computer-based systems

This standard is applicable to computer system hardware for
systems of Class 1 and 2 (as defined by IEC 61513 [III–5]) in NPPs.
This edition reflects recent developments in computer system
hardware design, the use of predeveloped hardware and changes in
terminology.

IEC 62138
[III–9]

Nuclear power plants — instrumentation and control systems
important to safety — software aspects for computer-based systems
performing Category B or C functions

This standard specifies requirements for the software of
computer based I&C systems performing functions of safety
Category B or C, as defined by IEC 61226 [III–7]. It complements
IEC 60880, which provides requirements for the software of
computer based I&C systems performing functions of safety
Category A. It is consistent with, and complementary to, IEC 61513.
Activities that are mainly system level activities (e.g. integration,
validation, installation) are not addressed exhaustively by this
document; requirements that are not specific to software are
deferred to IEC 61513.

TABLE III–1. INTERNATIONAL STANDARDS THAT HAVE A STRONG RELATIONSHIP WITH THIS SAFETY REPORT (cont.)

IEC 62566 [III–10]	Nuclear power plants — instrumentation and control important to safety — development of HDL-programmed integrated circuits for systems performing Category A functions
	This international standard provides requirements for achieving highly reliable hardware description language (HDL) programmed devices (HPDs for use in I&C systems of NPPs performing functions of safety Category A, as defined by IEC 61226 [III–7]. The programming of HPDs relies on HDL and related software tools. They are typically based on blank field programmable gate arrays or similar microelectronic technologies. General purpose integrated circuits such as microprocessors are not HPDs.
IEEE 7-4.3.2 [III–11]	IEEE Standard criteria for programmable digital devices in safety systems of nuclear power generating stations
	This standard provides additional specific requirements to supplement the criteria and requirements of IEEE 603 [III–12], which are specified for programmable digital devices. Within the context of this standard, a programmable digital device is any device that relies on software instructions or programmable logic to accomplish a function. Examples include a computer, a programmable hardware device and a device with firmware. Systems using these devices are also referred to as digital safety systems in this standard. The criteria contained therein, in conjunction with criteria in IEEE 603 [III–12], establish the minimum functional and design requirements for programmable digital devices used as components of a safety system.

TABLE III–1. INTERNATIONAL STANDARDS THAT HAVE A STRONG RELATIONSHIP WITH THIS SAFETY REPORT (cont.)

IEEE 1012 [III–13]	IEEE Standard for system, software, and hardware verification and validation
	Verification and validation (V&V) processes are used to determine whether the development products of a given activity conform to the requirements of that activity and whether the product satisfies its intended use and user needs. V&V life cycle process requirements are specified for different integrity levels. The scope of V&V processes encompasses systems, software and hardware, and it includes their interfaces. This standard applies to systems, software and hardware that are developed, maintained or reused (legacy, commercial off the shelf, non-developmental items). The term 'software' also includes firmware and microcode. The system, software and hardware should include documentation. V&V processes include the analysis, evaluation, review, inspection, assessment and testing of products.
IEEE 1074 [III–14]	IEEE Standard for developing a software project life cycle process
	This standard provides a procedure for creating a software project life cycle process. It is primarily directed at the process architect for a given software project. (Replaced by ISO/IEC TR 24774 [III–15].)

ᵃ Parts 1–4 are normative and parts 5–7 are informative. Certified products conform to parts 1/2/3/4, 1/2/4 or 1/3/4. Smart devices by definition conform to parts 1/2/3/4.

TABLE III–2. NATIONAL STANDARDS THAT HAVE A STRONG RELATIONSHIP WITH THIS SAFETY REPORT

RCC-E [III–16]	Design and construction rules for electrical and I&C systems and equipment
	This code describes the rules for designing and building electrical assemblies and I&C systems for pressurized water reactors. Volume III deals with automation and control systems, with sections III.H and III.I providing practical guidance for the use and qualification of smart devices in particular (referred to as DDLFs). It offers two paths for qualifying DDLFs: one with the use of IEC 62671 [III–1] and one for IEC 61508 [III–4] precertified devices. It lays out the methodology for applying the standards and, in the case of pre-existing certifications, provides a framework for making use of existing work, supplemented with an audit (graded according to safety class). It also includes provisions for compensating measures in case of gaps.

TABLE III–2. NATIONAL STANDARDS THAT HAVE A STRONG RELATIONSHIP WITH THIS SAFETY REPORT (cont.)

NS-TAST-GD-046 Revision 6 [III–17]	Computer based safety systems — nuclear safety technical assessment guide
	The purpose of this Technical Assessment Guide is to provide additional guidance for applying safety assessment principle ESS.27 [III–18], which presents the elements of a multipart procedure that should be used to demonstrate the adequacy of a computer based safety system. It expands upon the guidance provided by ESS.27 to assist Office for Nuclear Regulation assessors in applying judgement when assessing the adequacy of safety cases for computer based safety systems.
NRC Regulatory Issue Summary 2016-05 [III–19]	Embedded digital devices in safety related systems
	The intention of the United States Nuclear Regulatory Commission (NRC) in issuing this Regulatory Issue Summary was to heighten awareness that embedded digital devices might exist in procured equipment used in safety related systems without the devices having been explicitly identified in procurement documentation. Inadequate consideration of these devices in digital technology system upgrades, component replacements and new equipment applications could lead to an adverse safety consequence. Therefore, addressees should implement early efforts to identify these devices.
NRC Regulatory Issue Summary 2002-22, Supplement 1 [III–20]	Clarification on endorsement of nuclear energy institute guidance in designing digital upgrades in instrumentation and control systems
	The guidance in this Regulatory Issue Summary supplement clarifies the NRC's endorsement of the guidance pertaining to NEI 01-01, sections 4 and 5 and appendices A and B. This supplement clarifies the guidance for preparing and documenting 'qualitative assessments' that can be used to evaluate the likelihood of failure of a proposed digital modification, including the likelihood of failure because of a CCF.
CSA N290.14-15 (R2020) [III–21]	Qualification of digital hardware and software for use in instrumentation and control applications for nuclear power plants
	This standard defines requirements for the process of qualification of digital hardware and software for use in I&C applications for NPPs. It applies to individual safety related programmable digital devices containing software or programmable logic (e.g. devices such as application specific integrated circuits, complex programmable logic devices and field programmable gate arrays).

REFERENCES TO ANNEX III

[III–1] INTERNATIONAL ELECTROTECHNICAL COMMISSION, Nuclear Power Plants, Instrumentation and Control Important to Safety, Selection and Use of Industrial Digital Devices of Limited Functionality, IEC 62671, IEC, Geneva (2013).

[III–2] INTERNATIONAL ELECTROTECHNICAL COMMISSION, Functional Safety of Electrical/Electronic/Programmable Electronic Safety-Related Systems, IEC 61508, IEC, Geneva (2010).

[III–3] INTERNATIONAL ELECTROTECHNICAL COMMISSION, Nuclear Power Plants, Instrumentation and Controls Important to Safety, Selection and Use of Industrial Digital Devices of Limited Functionality, IEC 62671, IEC, Geneva (2013); Corrigendum IEC 62671/COR1 (2016).

[III–4] INTERNATIONAL ELECTROTECHNICAL COMMISSION, Functional Safety of Electrical/Electronic/Programmable Electronic Safety-Related Systems, IEC 61508, IEC, Geneva (2010).

[III–5] INTERNATIONAL ELECTROTECHNICAL COMMISSION, Nuclear Power Plants, Instrumentation and Control Important to Safety, General Requirements for Systems, IEC 61513, IEC, Geneva (2016).

[III–6] INTERNATIONAL ELECTROTECHNICAL COMMISSION, Nuclear Power Plants, Instrumentation and Control Systems Important to Safety, Software Aspects for Computer-Based Systems Performing Category A Functions, IEC 60880, IEC, Geneva (2006).

[III–7] INTERNATIONAL ELECTROTECHNICAL COMMISSION, Nuclear Power Plants, Instrumentation and Control Systems Important to Safety, Categorization of Functions and Classification of Systems, IEC 61226, IEC, Geneva (2020).

[III–8] INTERNATIONAL ELECTROTECHNICAL COMMISSION, Nuclear Power Plants, Instrumentation and Control Systems Important to Safety, Hardware Requirements, IEC 60987, IEC, Geneva (2021).

[III–9] INTERNATIONAL ELECTROTECHNICAL COMMISSION, Nuclear Power Plants, Instrumentation and Control Systems Important to Safety, Software Aspects for Computer-Based Systems Performing Category B or C Functions, IEC 62138, IEC, Geneva (2018).

[III–10] INTERNATIONAL ELECTROTECHNICAL COMMISSION, Nuclear Power Plants, Instrumentation and Control Important to Safety, Development of HDL-Programmed Integrated Circuits for Systems Performing Category A Functions, IEC 62566, IEC, Geneva (2012).

[III–11] INSTITUTE OF ELECTRICAL AND ELECTRONIC ENGINEERS, IEEE Standard Criteria for Programmable Digital Devices in Safety Systems of Nuclear Power Generating Stations, IEEE 7-4.3.2, IEEE, New York (2016).

[III–12] INSTITUTE OF ELECTRICAL AND ELECTRONIC ENGINEERS, IEEE Standard Criteria for Safety Systems for Nuclear Power Generating Stations, IEEE 603, IEEE, New York (2018).

[III–13] INSTITUTE OF ELECTRICAL AND ELECTRONIC ENGINEERS, IEEE Standard for System, Software, and Hardware Verification and Validation, IEEE 1012, IEEE, New York (2016).

[III–14] INSTITUTE OF ELECTRICAL AND ELECTRONIC ENGINEERS, IEEE Standard for Developing a Software Project Life Cycle Process, IEEE 1074, IEEE, New York (2006).

[III–15] INTERNATIONAL ORGANIZATION FOR STANDARDIZATION, INTERNATIONAL ELECTROTECHNICAL COMMISSION, INSTITUTE OF ELECTRICAL AND ELECTRONIC ENGINEERS, Systems and software engineering, Life cycle management, Specification for process description, ISO/IEC/IEEE 24774, ISO/IEC & IEEE, Geneva (2021).

[III–16] FRENCH ASSOCIATION FOR DESIGN, CONSTRUCTION AND IN-SERVICE INSPECTION RULES FOR NUCLEAR ISLAND COMPONENTS, Design and Construction Rules for Electrical and I&C Systems and Equipment, RCC-E, AFCEN, Paris (2016).

[III–17] OFFICE FOR NUCLEAR REGULATION, Computer Based Safety Systems, NS-TAST-GD-046 Revision 6, CM9 Folder 1.1.3.978 (2020/261582), ONR, Bootle (2019),
http://www.onr.org.uk/operational/tech_asst_guides/ns-tast-gd-046.pdf

[III–18] OFFICE FOR NUCLEAR REGULATION, Safety Assessment Principles for Nuclear Facilities, 2014 Edition, Revision 1 (January 2020), ONR CM9 Ref 2019/367414, ONR, Bootle (2020).

[III–19] NUCLEAR REGULATORY COMMISSION, Embedded Digital Devices in Safety-Related Systems, NRC Regulatory Issue Summary 2016-05, NRC, Washington, DC (2016).

[III–20] NUCLEAR REGULATORY COMMISSION, Clarification on Endorsement of Nuclear Energy Institute Guidance in Designing Digital Upgrades in Instrumentation and Control Systems, NRC Regulatory Issue Summary 2002-22, Supplement 1, NRC, Washington, DC (2018).

[III–21] CANADIAN STANDARDS ASSOCIATION, Qualification of Digital Hardware and Software for Use in Instrumentation and Control Applications for Nuclear Power Plants, CSA N290.14-15 (R2020), CSA Group, Toronto (2020).

Annex IV

EXAMPLES OF MEMBER STATE PRACTICES

IV–1. FRAMEWORK FOR QUALIFICATION OF SMART DEVICES (FRANCE)

The RCC-E standard [IV–1] defines the framework for smart device qualification. All programmable digital devices need to undergo software qualification (referred to simply as 'qualification' hereafter in this annex) in addition to the hardware qualification process.

Programmable devices that meet the definition of DDLFs according to IEC 62671 [IV–2] may be qualified using a simpler mechanism than that required for computer based systems (i.e. the use of nuclear standards such as IEC 62138 [IV–3], IEC 60880 [IV–4] or IEC 62566 [IV–5]). In the case of DDLFs, two qualification paths are possible: for devices holding IEC 61508 [IV–5] certification, qualification can be obtained on the basis of the existing certification and an additional audit; for others, the IEC 62671 [IV–2] standard can be used.

IV–1.1. Smart device already certified according to IEC 61508

If the device was developed and certified according to the requirements of IEC 61508 [IV–6], RCC-E [IV–1] allows making use of existing work. In this case, the qualification effort will be limited to the submission of the certification report and hosting an audit that is graded according to the safety class. The end result of the audit is the evaluation and application report.

Before the audit, a device suitability assessment is carried out to ensure that the device meets the criteria for this qualification pathway. A minimum of SIL 1, 2 or 3 is required for Class 3, 2 and 1 evaluation, respectively. The certified safety function must be compatible with the intended use of the device.

The audit involves the inspection of the following themes: life cycle, software architecture, verification activities, exercise in traceability of requirements (down to the source code), demonstration of proficiency in the use of tools, inspection of validation test bench, inspection of static analysis measures applied and inspection of defensive programming measures employed.

If the device is to be modified in the future, the certification needs to be updated, and an impact analysis must be carried out to demonstrate that the modification does not fundamentally change the device.

IV–1.2. Digital devices with limited functionality according to IEC 62671

RCC-E [IV–1] uses the IEC 62671 [IV–2] standard for the qualification of DDLFs; RCC-E is a reference for use with the standard. The RCC-E framework defines the qualification process for a generic, rather than a specific, application treated in the standard. The candidate selection process defined in the standard is not used. Instead, the selection (as well as integration of the device) is performed within the life cycle of the system, following IEC 61513 [IV–7].

IEC 62671 is used except for clauses 5 and 8. The principles of clause 5 of IEC 62671 are given in RCC-E [IV–1], so only the definition of a DDLF is used. In addition, IEC 61513 [IV–7] is the reference for integrating pre-existing components into a system, so it replaces IEC 62671, clause 8, 'Integration into the application'. The rest of the standard is applied by RCC-E, and the qualification is reported in the evaluation and application report as defined in the IEC 62671 [IV–2] standard.

IV–1.3. Checking compliance with standards

There are several ways to verify development quality, as detailed in subsections IV–1.3.1 to IV–1.3.3.

IV–1.3.1. Clause by clause compliance analysis

A typical Member State practice is for the licensee to be responsible for equipment compliance. Third party certification is thus not normally accepted as the sole guarantee of compliance with standards (although it can be used as additional evidence). Typically, the licensee performs a clause by clause analysis to ensure compliance with the relevant standard.

The clause by clause analysis normally takes the form of a conformity matrix or safety case. This matrix tends to contain at least the following elements:

— On each line, one elementary requirement with one 'shall' or one 'should';
— An evaluation of conformity for each clause with elementary requirements, such as non-compliant, partly compliant, compliant or not applicable;
— For each clause, a justification to support the conformity state, with references to supplier documentation as evidence.

IV–1.3.2. Audit graded by safety class

In some cases, for precertified devices, an audit might be a suitable way to verify compliance. The audit usually involves the examination of the full report of the independent certification body, together with the manufacturer's documentation used as evidence for standard compliance. Code review, as well as some additional testing, may be used, and its extent is graded according to the safety class of the device.

IV–1.3.3. Compensatory measures

In the case of non-conformities, there might be a need to demonstrate that the same objective is achieved using different means, such as complementary testing or code analysis (e.g. code reviews, static analysis, statistical testing), and preparation of additional documentation. IEC 62671 [IV–2] identifies suitable compensatory measures as a function of the application class and the type of evidence.

IV–2. FRAMEWORK FOR QUALIFICATION OF SMART DEVICES (UNITED KINGDOM)

The general approach adopted in the United Kingdom (UK) for the qualification of smart devices aligns with the practice described in NR-T-3.31 [IV–8] and this report.

IV–2.1. General context related to smart device qualification

The key principle that underpins the approach to qualification of smart devices for use in nuclear safety applications is based on the regulatory framework in the UK as set out in relevant safety assessment principles [IV–9] developed by the Office for Nuclear Regulation (ONR) in conjunction with nuclear site licensees and other stakeholders. The key safety assessment principle applicable to smart devices, ESS.27 [IV–9], states:

"Where the system reliability is significantly dependent upon the performance of computer software, compliance with appropriate standards and practices throughout the software development lifecycle should be established in order to provide assurance of the final design."

According to NR-T-3.31 [IV–8], the additional guidance supporting this principle, including ONR Technical Assessment Guide 46 [IV–10], "clarifies that, because of the complexity of these devices, traditional methods of reliability assessment are typically not sufficient to manage the risk of systematic failures, and additional activities are expected as part of their justification." This is normally achieved by means of a two pronged approach to demonstrate the suitability of a smart device for a UK nuclear safety application. The two prongs consist of the following [IV–8]:

— Production excellence, which focuses on the demonstration of excellence in all aspects of production, from the initial specification through to commissioning a smart device into operational service;
— Independent confidence building measures (ICBMs), which provide an independent and thorough assessment of a smart device's fitness for purpose in a nuclear safety application.

Additional information on the interpretation of this principle for smart device qualification (including an overview of the UK regulatory framework) is provided in annex IV of Ref. [IV–8].

IV–2.2. 'Emphasis' approach to assessing production excellence of a smart device

The use of a methodology referred to as 'Emphasis' [IV–11] has been adopted by UK nuclear site licensees for assessing the production excellence of a smart device in relation to the quality of the manufacturer's development life cycle (see also Ref. [IV–11]). This approach was developed in the UK by the Control and Instrumentation Nuclear Industry Forum (CINIF) on the basis of the clauses of IEC 61508 [IV–6] that are relevant to smart devices. Emphasis contains a set of approximately 300 questions that are graded to the safety class of the intended application. The questions are subdivided into 'phases', and examples are given below.

Phase 1: QA and safety management

— Does your company monitor the field performance of its products in any way?
— Is a configuration control system operated that uses version control for all design documentation, hardware, firmware and software?
— Were all items, including hardware and firmware design information, placed under formal configuration control before commencement of any formal verification activity or validation phase in which they were used?

Phase 2: generic programmable electronic aspects and development process for the device as a whole

— Was a specification document produced and agreed for this product before development work started?
— If a modification was made during development that pertains to an earlier phase, was an impact analysis performed to determine which components were affected and which earlier activities are repeated?
— Did the validation activities demonstrate that all required functions and performance requirements of the product were correctly implemented?

Phase 3: hardware development process and verification activities

— Was the output of each design and development phase verified for conformance to the hardware specification?
— Was the (hardware) specification subject to an inspection process?
— Was functional testing performed under the full range of tolerable environmental conditions?

Phase 4: software development process and verification activities

— Was verification performed and were results documented for each phase of the software life cycle?
— Does the design ensure that response times are predictable and consistent?
— Were structured diagrammatic methods used during software architecture design and detailed design and coding?
— Were test cases derived from boundary values and, by partitioning the inputs into equivalence classes, used for system validation?
— Was control flow analysis used during software verification?

Although Emphasis covers some basic security aspects, the expectation is that this will be complemented with other activities once the target application is confirmed.

Emphasis addresses the aspects related to the quality of the manufacturer's product development life cycle. A full production excellence assessment may require additional information, such as hardware reliability assessments and other hardware assessments (e.g. environmental assessments).

Application of the Emphasis methodology is supported by a dedicated tool developed and maintained by CINIF. The assessment of production excellence based upon use of the Emphasis tool enables collaborative working between the assessors and the smart device vendor. The aim of this activity is for the assessor to determine whether the expectations relating to each question are satisfied (including recording the evidence to support that judgement) or whether additional compensating activities are needed to address a gap. The tool also allows relevant evidence needed to support the qualification to be integrated as part of the supporting information.

In the UK context, the safety justification is structured according to a claim, argument and evidence approach, which means that any statements and the reasoning leading to them need to be underpinned by evidence in order to demonstrate the following:

— The credibility of the claim;
— Adequate application of any technique that might be used as part of the argument.

In the context of production excellence assessment, evidence can take the form of, for example, testing results, documents, analysis results or other artefacts, and such evidence can support qualitative or quantitative arguments.

An example of the graphical interface of the Emphasis tool is presented in Fig. IV–1 to highlight the following:

— The different steps in the assessment process (e.g. the 'answer' tab used by the manufacturer and the 'evaluation' tab used by the assessor);
— The evidence expected to be provided to support a claim made in the answer;
— The expert judgement made on the evidence on the 'adequacy' or 'gap' (which then needs to be resolved with a compensating activity).

The Emphasis process and tool aim at capturing some of the logic behind the techniques and measures tables of IEC 61508 [IV–6], including the levels of recommendation and levels of effectiveness for specific techniques based on the required SIL, as shown in Fig. IV–2.

Emphasis is an established approach among UK nuclear licensees to support meeting production excellence requirements, but there are cases in which completing an Emphasis assessment may prove difficult because of issues such as unavailability of evidence or lack of manufacturer support. In such cases, if product qualification is to be pursued, an alternative approach to support the demonstration of production excellence needs to be sought and justified. Section IV–2.3 expands on such an alternative approach.

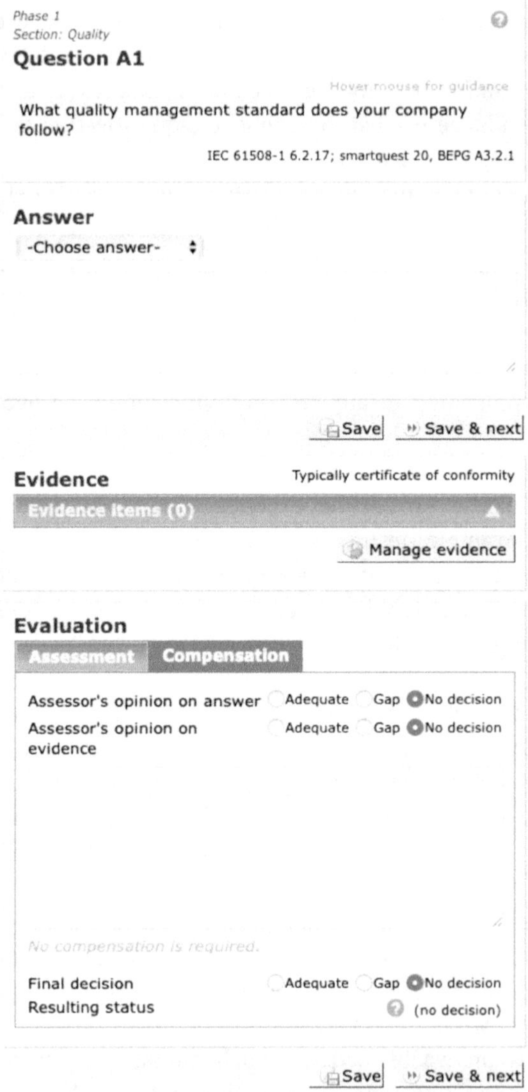

FIG. IV–1. Example of the graphical interface of the Emphasis tool, showing the answer, evidence and assessment tabs [IV–11] (figure courtesy of CINIF).

IV–2.3. Alternative approach to support smart device qualification

An alternative approach to support the production excellence of a device was developed by CINIF. This method, which is generally referred to as 'Cogs', as explained in Ref. [IV–12], is based on a claim, argument and evidence approach.

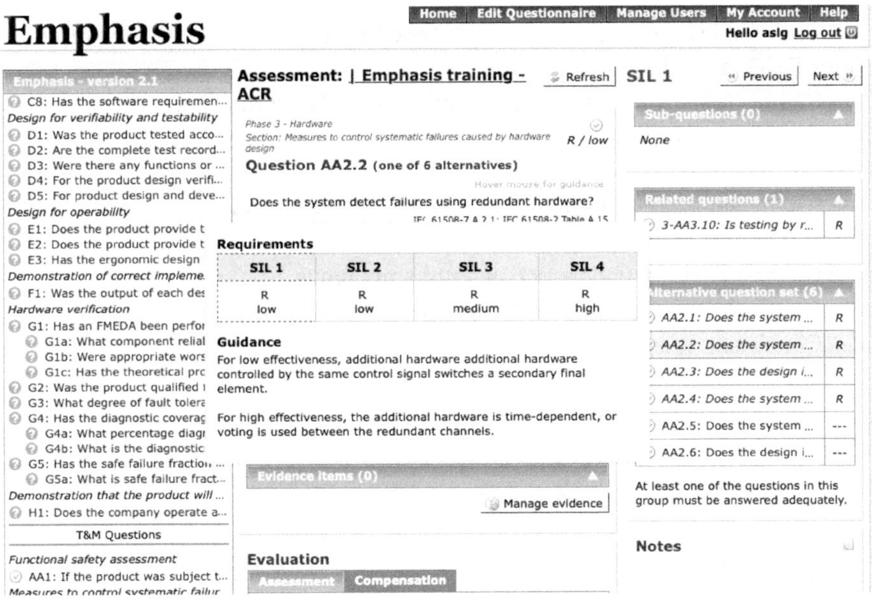

FIG. IV–2. Interface to the Emphasis tool, with the questions bar and pop-up box, showing the levels of recommendation for a specific technique or measure (figure courtesy of CINIF).

While its application in the UK nuclear industry is currently more limited than that of Emphasis, and while it is not the preferred route to support a smart device qualification, Cogs — if suitably implemented — has the potential to provide adequate confidence in the production excellence of a device. The Cogs approach can also be employed to identify suitable compensatory activities if significant gaps are encountered within an Emphasis assessment (or in other parts of a production excellence assessment) for which a resolution is not straightforward. The Cogs approach can be tailored to capture the claim, argument and evidence structure of interest and is not limited to the product development life cycle. Thus, it has the potential to support the full scope of PE, including, for example, hardware assessments.

At high level, the Cogs approach consists of a set of top level claims to be justified and is supported by guidance on how to expand and justify these claims. The top level claims in Cogs are the following:

— Claim 1: the behaviour and functionality are adequately documented;
— Claim 2: the device behaves according to documentation (initially);
— Claim 3: the device carries on behaving according to documentation;
— Claim 4: sound development process and design principles are used in the device development.

Each top level claim is then decomposed in subclaims, which are presented in Figs IV–3 to IV–6.

Figure IV–3 presents how Claim 1 can be further decomposed into subclaims, in terms of the device behaviour being as follows:

— Complete: it does not leave out any aspects of the behaviour that might be relevant to an application;
— Clear: it is easily understood and not ambiguous;
— Verifiable: it makes assertions that are possible to test;
— Consistent: it is not contradictory.

Completeness can be further developed by an attribute decomposition, breaking it down into a number of relevant behavioural properties, such as the following examples:

— Functionality, namely the functional behaviour of the device, including all the functions and operating modes;
— Performance, namely the characteristics defining the ability to achieve the intended functions, such as accuracy, time response and throughput;
— Dependability, including reliability, maintainability, failure integrity and security.

Once an adequate description of the behaviour is achieved (Claim 1), the main task in Claim 2 is to confirm that the behaviour actually implements this description, showing the following:

— The behaviour is as described if the needs of the device are satisfied;
— The behaviour is understood in the presence of postulated internal and external non-nominal conditions.

Claim 2 is decomposed into subclaims, as presented in Fig. IV–4, where, for each property, a claim is made that the behaviour achieved by the device is the same as that claimed in the description. Supporting evidence, typically from testing or operational experience together with knowledge of the instrument, is used to justify that each claim holds.

Claim 3 aims at confirming that after installation, the behaviour of the component will continue to be as described over its entire lifetime, in spite of the component's evolving environment, capabilities and any changes, be them deliberate, planned, accidental or out of the user's control. The consistent behaviour over the component's lifetime depends on the following:

— The correct environment of the component and the continuing fulfilment of the device's needs;
— The modifiability of the component and how likely faults are to be introduced when the component is modified, either via deliberate changes, such as calibration, or unintended changes, such as damage or vandalism;
— Changes resulting from ageing being rendered benign by corrective or preventative maintenance.

Figure IV–5 presents the main subclaims under Claim 3. While Claims 1–3 focus on the device and its behaviour, Claim 4 (Fig. IV–6) relates to the following:

— The development process and supporting processes, such as QA processes and configuration management;
— Design principles;
— Compliance with identified relevant standards.

This claim provides important support for the justification of the device behaviour presented in the previous claims, as processes such as configuration management and QA are crucial in the production of traceable and consistent evidence. Evidence that a sound development process was followed and that appropriate QA principles were used increases confidence in all the evidence generated, the relevance of the evidence in supporting the justification and, consequently, the overall justification. A high integrity development process makes use of design principles developed through experience and the accumulation of good practice. In general, these are expected to contribute to the reliability of the resulting device, as well as to the success of the overall development process.

While the claims are different from the qualification objectives identified in Section 4.1.2., there are similarities in the objectives and in the overall approach to the smart device qualification.

IV–2.4. Independent confidence building measures for smart devices

The ICBM element of ONR safety assessment principle ESS.27 [IV–9] is intended to confirm the adequacy of a smart device by independent means. This step is particularly important when considering commercial off the shelf smart devices that may not have been developed specifically for use in the nuclear sector. As for the production excellence element, the expectation is for the

ICBMs to be graded to safety classification. Examples of ICBMs are provided in ONR Technical Assessment Guide 46 [IV–10] and reproduced in Table IV–1.

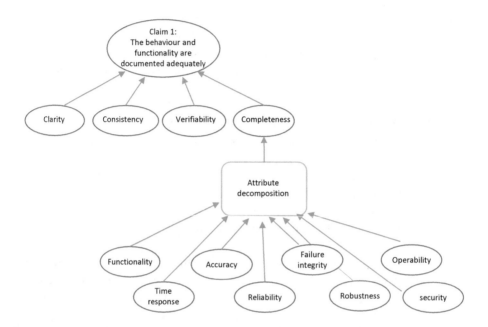

FIG. IV–3. Claim 1 in Cogs (figure reproduced from Ref. [IV–12] with permission).

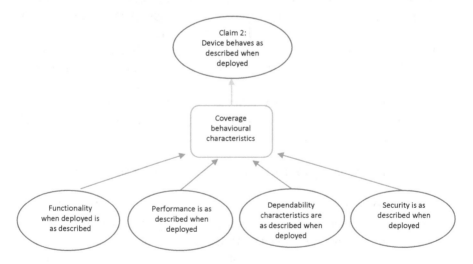

FIG. IV–4. Claim 2 in Cogs (figure reproduced from Ref. [IV–12] with permission).

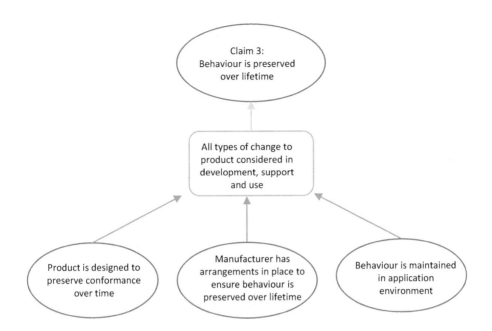

FIG. IV–5. Claim 3 in Cogs (figure reproduced from Ref. [IV–12] with permission).

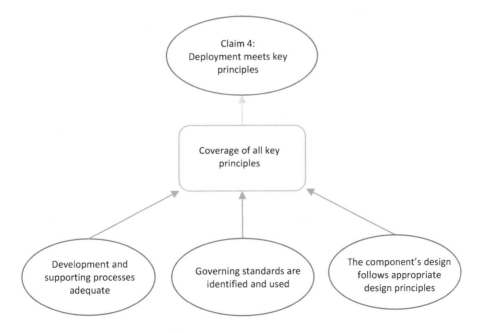

FIG. IV–6. Claim 4 in Cogs (figure reproduced from Ref. [IV–12] with permission).

TABLE IV–1. EXAMPLE GRADED APPROACH TO THE SELECTION OF MEASURES TO SUPPORT JUSTIFICATION OF COMMERCIAL OFF THE SHELF SMART DEVICES

(reproduced with permission from Ref. [IV–10])

	Class 3	Class 2	Class 1 (SIL 3 only)
Production excellence	There should be evidence of production excellence assessed using techniques and measures appropriate for SIL 1 Application of compensating activities as required to address gaps in production excellence	There should be evidence of production excellence assessed using techniques and measures appropriate for SIL 2 Application of compensating activities as required to address gaps in production excellence	There should be evidence of production excellence assessed using techniques and measures appropriate for SIL 3 Application of compensating activities as required to address gaps in production excellence

IV–2.5. Practical experience in smart device qualification

Licensees in the UK adhere to company specific guidance documents that, although developed using licensee expertise, aim at ensuring alignment with the goals enshrined in Ref. [IV–10]. The Emphasis approach discussed in Section IV–2.2 is common to all UK licensees. At Electricité de France Energy Nuclear Generation Limited (EDF NGL), for example, in addition to Emphasis, to assess the full scope of PE, an assessment of hardware reliability and a general hardware assessment are conducted (e.g. using manufacturer's type testing records and certificates of conformance as evidence), as well as an assessment of user documentation and the suitability of the production line. The choice of ICBMs is commensurate with Table IV–1. Substantial experience has been built up in the use of certain techniques, such as static code analysis and statistical testing, for the purpose of compensatory activities during a production excellence assessment or as an ICBM.

IV–2.5.1. Examples of techniques used as independent confidence building measures or compensatory activity

IV–2.5.1.1. Statistical testing

Within EDF NGL, statistical testing has been used successfully both as a compensatory activity and as an ICBM. In some cases it has identified the incorrect configuration of a device and in others it has detected an unknown feature of a smart device or a misinterpretation of the application environment. When a set of failure free statistical tests has been executed under a validated operational profile and under valid assumptions about device internal behaviour, the outcome can be used in support of a numerical claim on a probability of failure per demand (see also Ref. [IV–13]).

Statistical testing has been found to be a powerful technique to validate a device's behaviour under a simulation of actual plant application and is thus an effective means to support a safety case claim. Challenges encountered have revolved around the effort required to build test rigs and correctness checkers (Oracles). To address this challenge, the UK nuclear industry, through the aforementioned CINIF consortium, has funded the development of a generic smart device integrity test station for open loop statistical testing. This test station is intended to enable the testing of a variety of devices under various configurable operational profiles.

IV–2.5.1.2. Static code analysis

Static analysis techniques have been used by EDF NGL in a number of cases, usually as compensatory activity, and mostly in the form of tool based integrity checking static analysis. This has been found to add value, for example, where gaps in software design documentation or software verification documentation were found. In these cases, it can help to generate equivalent documentation and assurance.

It has also been found useful, but more difficult to apply successfully, when the development of the code itself (rather than the documentation thereof) was suboptimal. In such cases, static code analysis methods can generate a large number of findings that have to be sentenced. This requires substantial resources and support from the manufacturer, and can under certain circumstances require a change to the source code and subsequent reanalysis. Since it is frequently not possible to identify whether a detected issue will definitely lead to a failure at some point in time, or whether this is unlikely, some measure has to be taken to mitigate it.

It has been found that static code analysis, when applied with an appropriate level of rigour and expertise, has led to improvements in a manufacturer's code and code documentation. Appropriate rigour means that the techniques used need to be sufficiently probing, and the results need to be sentenced, taking into account the intended application context and other mitigating evidence in order to achieve a level of pragmatism.

Static code analysis used as an ICBM is useful where it adds techniques that are sufficiently diverse from those used by the manufacturer while still adding additional confidence in failure free operation of the device.

IV–2.5.1.3. Proven-in-use arguments

Data from manufacturer sales records and defect reporting, as well as information from use at other power stations, have been used within EDF NGL to support a qualification argument. Because of the difficulty in obtaining data and records of sufficient detail, however, this approach is mostly used in a qualitative manner to confirm that the device has no record that would undermine the claim made through a production excellence assessment. Research currently being carried out under the umbrella of CINIF is investigating whether a framework can be developed under which data from field use can be used and benefitted from in a more structured way.

IV–2.5.2. Equipment database

In the UK, a national database of smart device assessment information is being developed that serves as a platform for UK licensees to share information on smart devices that have been Emphasis assessed. This does not contain any detailed assessment information, because such information is usually controlled and is held in appropriate systems at each licensee's site. The national database contains high level information on devices assessed and assessment targets achieved (e.g. class, SIL). This serves as a basis to identify opportunities to share assessments among licensees, thus reducing assessment effort as well as encouraging manufacturer buy-in through displaying a wider community interest in Emphasis assessed devices.

IV–2.5.3. Intelligent customer role

Although UK licensees widely use external expert contractors to carry out assessments, the licensees themselves become involved in the assessment as 'intelligent customers', thus obtaining a level of oversight and influence on the assessment. The expectations for an 'intelligent customer' are set by ONR in Ref. [IV–14]. This ensures that decisions on deploying a smart device in a nuclear safety application can be made and owned by the licensees themselves.

An important part of the intelligent customer role is to arrive at a final judgement as to the suitability of a smart device for use in a specific NPP application. To this end, a judgement needs to be made as to the overall picture of evidence, gaps, mitigations and compensatory activities. It is noteworthy that it is not expected to achieve 100% compliance with every clause in Emphasis; rather, the overall picture has to be satisfactory. Furthermore, rather than looking to meet the exact requirement expressed in a clause, the spirit of the clause needs to be addressed.

IV–2.5.4. Summary

In summary, the Emphasis approach has been found to be beneficial in support of a production excellence assessment, as it uses a targeted question set based on IEC 61508 [IV–6] and is closely aligned with this standard but is tailored to commercial off the shelf smart devices. The Emphasis tool enables a collaborative approach among manufacturers, assessors and licensees. The approach is sound and has been used and further developed over more than a decade. It is being maintained and subjected to periodic review under the umbrella of CINIF. It is an approach shared by all UK licensees and is in alignment with regulatory requirements. It is based on identifying and gathering evidence, which means that statements are underwritten by evidence. In some cases, it has helped manufacturers to improve their processes.

Involvement by the licensee as an intelligent customer has proved beneficial in ensuring that qualification activities are informed by actual safety case and plant system information, and in supporting the licensee to arrive at a final judgement on the acceptability of a smart device.

The use of ICBMs has been found to be very beneficial, especially when static analysis and intelligent test approaches are combined in an effective and efficient manner. Experience at EDF NGL suggests that a tailored combination of static analysis and statistical testing techniques is the most suitable approach to increase confidence in the failure free operation of a device. It requires expert input to determine which combination to use in a particular qualification context.

The current challenges of the approach described in this section include obtaining manufacturer buy-in, since it can be difficult to persuade manufacturers to support the effort required in answering Emphasis questions and hosting a 4–5 day site visit. A further challenge comes from the fact that for commercial off the shelf devices, the life cycle applied does not always strictly conform with IEC 61508 [IV–6] or provide equivalence to it. This is especially true for devices that are not seen as 'safety devices' or for smart devices that have developed over time from an analogue version of the device.

A large degree of interpretation of the Emphasis questions is required to decide whether sufficient evidence is available in support of the overall case that Emphasis is aiming at. It has also been found that compensatory activities such as static code analysis and statistical testing can find issues in the device, and these need to be rectified or sentenced as not impacting safety before the device qualification can be considered complete. This can add substantially to both effort and timescales.

Finally, although Emphasis is a very rigorous approach, items that have passed the qualification can in theory still fail in application. Therefore, the use of ICBMs needs to be intelligent and informed rather than purely a checklist approach. ICBMs need to be chosen to give optimal confidence that a smart device will not dangerously fail once employed in a specific application context.

REFERENCES TO ANNEX IV

[IV–1]   FRENCH ASSOCIATION FOR DESIGN, CONSTRUCTION AND IN-SERVICE INSPECTION RULES FOR NUCLEAR ISLAND COMPONENTS, Design and Construction Rules for Electrical and I&C Systems and Equipment, RCC-E, AFCEN, Paris (2016).

[IV–2]   INTERNATIONAL ELECTROTECHNICAL COMMISSION, Nuclear Power Plants, Instrumentation and Control Important to Safety, Selection and Use of Industrial Digital Devices of Limited Functionality, IEC 62671, IEC, Geneva (2013).

[IV–3]   INTERNATIONAL ELECTROTECHNICAL COMMISSION, Nuclear Power Plants, Instrumentation and Control Systems Important to Safety, Software Aspects for Computer-Based Systems Performing Category B or C Functions, IEC 62138, IEC, Geneva (2018).

[IV–4]   INTERNATIONAL ELECTROTECHNICAL COMMISSION, Nuclear Power Plants, Instrumentation and Control Systems Important to Safety, Software Aspects for Computer-Based Systems Performing Category A Functions, IEC 60880, IEC, Geneva (2006).

[IV–5]   INTERNATIONAL ELECTROTECHNICAL COMMISSION, Nuclear Power Plants, Instrumentation and Control Important to Safety, Development of HDL-Programmed Integrated Circuits for Systems Performing Category A Functions, IEC 62566, IEC, Geneva (2012).

[IV–6]   INTERNATIONAL ELECTROTECHNICAL COMMISSION, Functional Safety of Electrical/Electronic/Programmable Electronic Safety-Related Systems, IEC 61508, IEC, Geneva (2010).

[IV–7]   INTERNATIONAL ELECTROTECHNICAL COMMISSION, Nuclear Power Plants, Instrumentation and Control Important to Safety, General Requirements for Systems, IEC 61513, IEC, Geneva (2016).

[IV–8]   INTERNATIONAL ATOMIC ENERGY AGENCY, Challenges and Approaches for Selecting, Assessing and Qualifying Commercial Industrial Digital Instrumentation and Control Equipment for Use in Nuclear Power Plant Applications, IAEA Nuclear Energy Series No. NR-T-3.31, IAEA, Vienna (2020).

[IV–9]   OFFICE FOR NUCLEAR REGULATION, Safety Assessment Principles for Nuclear Facilities, 2014 Edition, Revision 1 (January 2020), ONR CM9 Ref 2019/367414, ONR, Bootle (2020).

[IV–10]   OFFICE FOR NUCLEAR REGULATION, Computer Based Safety Systems, NS-TAST-GD-046 Revision 6, CM9 Folder 1.1.3.978 (2020/261582), ONR, Bootle (2019),
          http://www.onr.org.uk/operational/tech_asst_guides/ns-tast-gd-046.pdf

[IV–11]   STOCKHAM, R., Emphasis on safety, E&T **2** (2009).

[IV–12]   GUERRA, S., CHOZOS, N., SHERIDAN, D., "Justifying digital COTS components when compliance cannot be demonstrated — the Cogs approach", Proc. Ninth Int. Topical Mtg on Nuclear Plant Instrumentation, Control, and Human–Machine Interface Technologies (NPIC&HMIT 2015), Charlotte, 2015, American Nuclear Society, La Grange Park, IL (2015).

[IV–13]   KUBALL, S., MAY, J., A discussion of statistical testing on a safety-related application, Proc. Inst. Mech. Eng. O **221** 2 (2007) 121–132.

[IV–14]   OFFICE FOR NUCLEAR REGULATION, Licensee Core Safety and Intelligent Customer Capabilities, NS-TAST-GD-049 Revision 7, CM9 Folder 1.1.3.978 (2020/265746), ONR, Bootle (2019),
          http://www.onr.org.uk/operational/tech_asst_guides/ns-tast-gd-049.pdf

DEFINITIONS

The following definitions apply for the purposes of this Safety Report.
Further definitions are provided in the IAEA Safety Glossary:
Terminology Used in Nuclear Safety and Radiation Protection (2018 Edition)
https://www.iaea.org/publications/11098/iaea-safety-glossary-2018-edition

compensatory evidence. Complementary information specific to the device that is appropriate to the intended application and other elements of evidence of correctness that directly address the requirements to be applicable to the device in question. Examples include evaluation of applicable and credible operational experience, verification of design outputs and statistical testing.

digital device of limited functionality. An industrial digital device that is available on the market and has the following characteristics: (a) it was not initially developed for nuclear application and contains software, including firmware, or software developed logic or programmed logic; (b) it is autonomous and performs only one conceptually simple principal function, which is defined by the manufacturer and is not modifiable by the user; (c) it is not designed to be reprogrammable; and (d) if it is reconfigurable, the configurability is limited to parameters related to compatibility with the process being monitored or controlled, or interfaces with connected equipment.

diversity[1]. The presence of two or more redundant systems or components to perform an identified function, where the different systems or components have different attributes, so as to reduce the possibility of common cause failure, including common mode failure.

Note 1: When the term 'diversity' is used with an additional attribute, the term diversity indicates the general meaning 'existence of two or more different ways or means of achieving a specified objective', while the attribute indicates the characteristics of the different ways applied, e.g. functional diversity, equipment diversity, signal diversity.

---

[1] Definition from INTERNATIONAL ATOMIC ENERGY AGENCY, Design of Instrumentation and Control Systems for Nuclear Power Plants, IAEA Safety Standards Series No. SSG-39, IAEA, Vienna (2016).

Note 2: See also the entry for 'functional diversity' in the IAEA Safety Glossary.

qualification of a smart device. A process that provides a degree of confidence that a smart device, including its software, hardware description language and process interfaces, performs its intended function during its service life in a manner commensurate with the importance to safety of the I&C system or component.

ABBREVIATIONS

AC	alternating current
CCF	common cause failure
CINIF	Control and Instrumentation Nuclear Industry Forum
CSA	Canadian Standards Association
DC	direct current
DDLF	digital device of limited functionality
EDF	Electricité de France
EDF NGL	EDF Energy Nuclear Generation Limited
EMI	electromagnetic interference
EPRI	Electric Power Research Institute
FMEA	failure mode and effects analysis
FPGA	field programmable gate array
HDL	hardware description language
HPD	HDL programmed device
I&C	instrumentation and control
ICBM	independent confidence building measures
IEC	International Electrotechnical Commission
IEEE	Institute of Electronic and Electrical Engineers
NCFSI	non-conforming, counterfeit, fraudulent and suspect items
NPP	nuclear power plant
ONR	Office for Nuclear Regulation
PE	production excellence
QA	quality assurance
RCC-E	French electrical design standard
RFI	radiofrequency interference
SIL	safety integrity level
UK	United Kingdom
V&V	verification and validation

CONTRIBUTORS TO DRAFTING AND REVIEW

Burzynski, M.	Sunport, United States of America
Cossinet, T.	Electricité de France, France
Duchac, A.	International Atomic Energy Agency
Fournier, R.D.	Exida LLC, Canada
Guerra, S.	Adelard LLP, United Kingdom
Kuball, S.	Electricité de France Energy Nuclear Generation Limited, United Kingdom
Lamb, C.	Sandia National Laboratory, United States of America
Overling, E.	Electricité de France, France
Park, G.	Korea Institute of Nuclear Safety, Republic of Korea
Picca, P.	Office for Nuclear Regulation, United Kingdom
Samek, M.	ČEZ, a.s., Czech Republic
Tienes, M.	Framatome, Germany
Watanabe, N.	Nuclear Regulation Authority, Japan
Zhao, J.	Nuclear Regulatory Commission, United States of America

Technical Meeting

Vienna, Austria: 17–21 February 2020

Consultants Meetings

Vienna, Austria: 25–29 March 2019, 4–8 November 2019, 17 November 2020

IAEA
International Atomic Energy Agency

ORDERING LOCALLY

IAEA priced publications may be purchased from the sources listed below or from major local booksellers.

Orders for unpriced publications should be made directly to the IAEA. The contact details are given at the end of this list.

NORTH AMERICA

Bernan / Rowman & Littlefield
15250 NBN Way, Blue Ridge Summit, PA 17214, USA
Telephone: +1 800 462 6420 • Fax: +1 800 338 4550
Email: orders@rowman.com • Web site: www.rowman.com/bernan

REST OF WORLD

Please contact your preferred local supplier, or our lead distributor:

Eurospan Group
Gray's Inn House
127 Clerkenwell Road
London EC1R 5DB
United Kingdom

Trade orders and enquiries:
Telephone: +44 (0)176 760 4972 • Fax: +44 (0)176 760 1640
Email: eurospan@turpin-distribution.com

Individual orders:
www.eurospanbookstore.com/iaea

For further information:
Telephone: +44 (0)207 240 0856 • Fax: +44 (0)207 379 0609
Email: info@eurospangroup.com • Web site: www.eurospangroup.com

Orders for both priced and unpriced publications may be addressed directly to:
Marketing and Sales Unit
International Atomic Energy Agency
Vienna International Centre, PO Box 100, 1400 Vienna, Austria
Telephone: +43 1 2600 22529 or 22530 • Fax: +43 1 26007 22529
Email: sales.publications@iaea.org • Web site: www.iaea.org/publications